Self-Healing Systems and Wireless Networks Management

Self-Healing Systems and Wireless Networks Management

Junaid Ahsenali Chaudhry

CRC Press
Taylor & Francis Group
Boca Raton London New York

CRC Press is an imprint of the
Taylor & Francis Group, an **informa** business

CRC Press
Taylor & Francis Group
6000 Broken Sound Parkway NW, Suite 300
Boca Raton, FL 33487-2742

First issued in paperback 2019

ISBN-13: 978-1-4665-5648-5 (hbk)
ISBN-13: 978-0-367-37935-3 (pbk)

Library of Congress Cataloging-in-Publication Data

Chaudhry, Junaid Ahsenali.
  Self-healing systems and wireless networks management / Junaid Ahsenali Chaudhry.
    pages cm
  Summary: "A self-healing network has a network architecture that can withstand a failure in its transmission paths. The biggest issue, to date, in self-healing systems is fault identification and its classification. This book uses the techniques of casual reasoning to classify the candidate faults from the fault identification process. It employs a similarity matrix in order to match the user activity log and its pattern in a transformed space. I then describes how to embed the self-healing policy, so the if the client has more faults related to the previous one, they can be dealt with at the client side"-- Provided by publisher.
  Includes bibliographical references and index.
  ISBN 978-1-4665-5648-5 (hardback)
  1. Wireless communication systems--Quality control. 2. Computer networks--Management. 3. Self-organizing systems. 4. Fault location (Engineering) I. Title.

TK5103.2.C45156 2013
004.6--dc23   2013014422

Visit the Taylor & Francis Web site at
http://www.taylorandfrancis.com

and the CRC Press Web site at
http://www.crcpress.com

Contents

Foreword

Absolute security is something that is absolutely nonexistent. In information security business, all we can do is to try to delay inevitable, unfortunate incidents that are going to cause disruption in the normal course of action in business systems. In a world where information availability is the most critical, the most common fear for the system managers is when either information is lost or corrupted or the networks, carrying that information, are attacked. The attacker's objective is the same: preventing information from reaching its intended audience.

System administrators and IT workers strive day in and day out to keep the connection between authentic data source and its intended audience. However, the threats to this front are increasing at a tremendous rate. The level of creativity among those threats is astonishingly high. For the IT workforce, policing the network and database boarders requires help that is more intelligent than old-style decision support systems and more vigilant than plain heuristically restricted expert systems. The advent of Self-healing Systems and their capabilities, the support for distributed intelligence, elimination of single point of failure, lightweightness, and robustness are some traits of the self-healing systems that are precisely what industry is seeking in a secure application framework.

Dr. Junaid Chaudhry has been developing smart solutions for the distributed networks since 2003. This book is the unification of his work that he has combined into a flexible Self-Healing Engine. During his stay at the National Center of

Excellence in Ubiquitous Systems and Networks, he worked on self-growing middleware that learns from the attacks to the infrastructure and, in fact, he was one of the foremost researchers who presented the idea of active policies and autonomic security management. The Self-Healing Engine also serves as a Unified Threat Management (UTM) mechanism for business systems; i.e., the Korean banking infrastructure that adopted his solution during the last decade. Currently, Dr. Chaudhry is leading our Research and Development team at the Information Security Center for law enforcement purposes.

His extensive experience of building up capability and mission critical systems in developing infrastructure makes him an ideal candidate for the job at hand: the development of cyber defense and unified threat management capability. His idea of extendible self-healing engine is so customizable that a small organization can even adjust it to their requirements and set up a small Security Operation Center (SOC) for themselves. Such capability is hard to find in off-the-shelf software products. I find his book to be a good read to the security managers and engineers alike and highly recommend it to everyone for a good read.

Captain Mahmod Ibrahim

Chief
Information Security Center,
Ministry of Interior,
State of Qatar

Preface

The biggest issue, to date, in self-healing systems is fault identification and its classification; that is, to identify what is a fault and what is not? What kind of fault is it? We use the techniques of Casual Reasoning in order to classify the candidate faults from the fault identification process. We devise a similarity matrix in order to match the user activity log and its pattern in a transformed space. The establishment and usage of a transformed space gives us scalability and eliminates the "out-of-context" fault candidates. Once the fault candidates are recognized, they are classified into an order and forwarded to the policy engine that was developed as a part of this research.

We observe that the software specifications of a software program remain constant throughout the lifecycle of software. We exploit the software specifications to generate XML format of the specifications and further convert them into XML trees. These tree structures assist us in their manipulation so that we could identify the patterns related to the fault notifications, which are forwarded through client, with the software specifications. Any abnormalities are enlisted in order to be forwarded to the casual reasoning section. The casual reasoning section identifies the faults and ranks the extracted regions. Every classification class has candidate matches in the plug-in bank that contains the executables related to certain faults.

After the operation of the casual reasoning section, the entire context is forwarded to the policy engine, which in collaboration with scheduler and plug-in bank generates a healing policy. The healing policy contains the solution of the fault and the faults related to that fault. We embed the solutions of the related faults in the healing policy so that if the client faces more faults related to the previous one, they could be dealt with at the client side.

The results obtained because of the experiments that we carried out show the evidence that the scheme proposed shows better performance than the schemes previously proposed. We fulfill the tasks in linear time, which means that the increase in the source file size does not affect the performance of our system. This makes our system highly scalable for distributed self-healing systems.

Chapter 1

Introduction

Motivation

To date, the natural growth path for computer systems has been in supporting technologies such as data storage density, processing capability, and per-user network bandwidth. The usefulness of Internet and intranet networks has fueled the growth of computing applications and in turn the complexity of their administration. In the presence of more than 30 billion pages on the "surface web," the "deep web" containing 450 to 500 times more volume in documents than the "surface web," data growth rate of approximately 217% and over 2.2 billion Internet users,[1] which is between 3 and 15 times larger than today's telephone system, the growth of information systems is astronomical. Furthermore, as the overall number of sensor devices in the world is increasing constantly at a rapid pace, these data sets offer a foretaste of the type of numerical techniques needed to resolve the functional challenges that will be faced in the future.

Scenario

Karen starts her morning by reading her online news service subscription. When her car arrives, she switches to reading the news on her PDA that is equipped with a wide-area wireless connection. In the newspaper, she finds an advertisement for a new wireless equipped camera and calls her friend David to tell him about it. David's home entertainment system reduces the volume of the currently playing music as his phone rings. Karen begins telling David about the camera, and forwards him a copy of the advertisement that pops up on his home display. David's decision to buy the camera involves another set of services to be downloaded on his PDA. One acts as a shopping agent to verify that the price is the best possible and another verifies that the camera manufacturer has a dealer in that area. When the purchase of the camera is made, David's biometric identification ring authenticates the transaction.

His digital camera comes equipped with a short-range radio transceiver as well as a removable cartridge. The data watch he wears can communicate with the camera as well. The two devices inform each other of their protocols and requirements when they are first brought within range of each other. When he snaps a picture, the photo data is communicated to the watch, where it is held until a connection can be made to the rest of the world. David's watch acts as a personal storage device and as a gateway to other connections.

Eventually, David enters his home and brings his watch close to his mobile network terminal. At this point, the photo data from the wristwatch can be injected into the wider area network for the next step of its journey. The pictures find their way through the network to the photo album service. When the camera takes the snapshot, it included his encrypted personal ID (from the ring with biometric safeguards) in the data packets. Lower-cost data transfer options are always considered automatically based on David's past usage patterns and likely future locations (extracted from his schedule). For example, if

David is near his home (and likely headed there), his cellular phone will not place a call to transfer the data, but rather defer the transfer to happen at home where a cheaper Internet connection is available through the home's wireless network and network portal. The most complex aspect to this scenario is how the decisions are going to be made without human intervention and how decisions are changed per situation.

Economic Considerations

A practical system of this nature would inevitably consist of devices and networks manufactured by different vendors, conforming to different, possibly competing, standards. Therefore, at least from an economic point of view, it is natural to ask why different vendors would want their devices to interoperate with one another. Although this dissertation addresses the technical challenges posed by this heterogeneity, presuming the commercial viability of such a model, this section hints at the feasibility of such a system in the not so distant future. The computer industry differs from traditional manufacturing in that the costs associated with a given product decline over time due to cheaper implementation. The trend for performance, on the other hand, is relentlessly upward. Hence, the economics of the industry enables manufacturers to periodically introduce enhanced models with richer feature sets at an affordable price. The result has been a proliferation of devices embedded with appreciable computing power. The Firetide[1] Web site shows that 98% of the world's processors go into embedded devices and not desktop computers. One natural consequence of the growing complexity of the devices is the desire to connect them in order to achieve greater functionality, flexibility, and utility. This potential has certainly been realized by the computer industry in the last few years and an increasing number of efforts are under way to enable interoperation of consumer devices, especially in the context of home and office automation.

In addition to the older systems like X.10 and CEBus, devices conforming to newer systems like Jini, UpnP, and OSGi have emerged in the market, indicating the industry trend toward enabling a ubiquitous system. More interestingly, users have also shown keen interest in "programming" their devices as long as they can perform useful functions for them, for example, the downloadable mobile phone tunes and games industry is estimated at $1.3USD billion in the U.K. These trends indicate that in the future device manufacturers would instrument their devices to enable a ubiquitous system to provide greater flexibility, functionality, and utility of their devices.

Problem Statement

Preemptive measures have done little to cut down on network management costs, and resource requirements for management are increasing with network size and complexity.[2] Autonomic computing (AC) provides a cheaper solution for robust network management in vastly expanding networks in the form of self-management. Self-management is a tool through which performance of the computer systems can be optimized without human user intervention. Turing suggests that autonomic systems have exponential complexity, which can hamper the appropriate problem marking and raise the software cost.[3] Therefore, it is critical to provide an incremental, low cost, and time-efficient solution along with minimal maintenance cost. In distributed computing environments, the area of problem domain becomes vast and in order to cover this domain, a self-growing reasoning facility that gives a "quick fix" is required. The casual reasoning is a candidate that contains the capabilities that fulfill the requirements of the system under study. Casual reasoning can cover large exception bases and amalgamate growing exceptions with no effect on precision and accuracy.

Hybrid networks cater with high levels of Quality of Service (QoS), scalability, and dynamic service delivery requirements. The amplified utilization of hybrid networks, that is, ubiquitous-Zone-based (u-Zone) networks, has given greater importance to human resources, downtime, and user training costs. The u-Zone networks are the fusion of the cluster of hybrid Mobile Ad-hoc NETworks (MANETs) and high-speed mesh network backbones (Figure 1.1). They provide robust wireless connectivity to heterogeneous wireless devices and take less setup time. The clusters of hybrid networks feature heterogeneity, mobility, dynamic topologies, limited physical security, and limited survivability.[4] The mesh networks also provide high-speed feedback to the connected clusters. The applications of MANETs vary greatly from disaster and emergency response to entertainment and Internet connectivity to mobile users. The u-Zone networks contain a significantly vast variety of devices connected to them. It is not appropriate to address the problems of each category of devices individually. We need to have some general solutions that could entertain a certain set of devices. Moreover, the probability of a management solution made for one type of client being appropriate

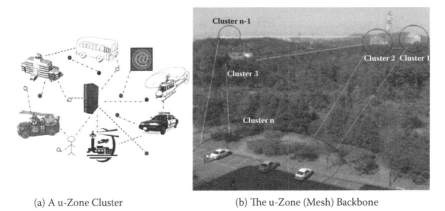

(a) A u-Zone Cluster   (b) The u-Zone (Mesh) Backbone

Figure 1.1   A zone-based network.

for another type is very low. Chaudhry and Park target self-management in hybrid environment through a "divide and conquer" approach by using component-based programming.[5] They propose to rapidly divide the problem into sub-domains and for each domain to then be assigned sub-solutions. The amalgamation of all sub-solutions gives the final management solution to the client devices. The following is a list of objectives and problems in hybrid networks:

- Huge management costs and need for unsupervised management and self-growing knowledge.
- Get benefits of ripple processing effect in diverse networks and facilitate proxy execution or even proxy management.
- Devise a novel reasoning facility and modify the structure of the conditional statement.
- Propose a clear hierarchy for autonomic management functions.
- Devise a novel reasoning facility and modify the structure of the conditional statement.
- Vision: If there are self-managed, self-growing networks, then why not self-managed, self-growing management systems?
- Logical complications and simplification of tasks.
- Biological example: Ali feels pain in his right index finger. Causes: 1. His finger is caught in a mouse trap. 2. Nail infection.
- Configuration: One puts his finger where he cannot see OR pathogens incubation period.
- Fault removal: Get his finger out of the trap OR external disinfectant.
- Healing: Psychological treatment OR immune system defense.
- Optimization: Put some gloves on OR patience and forbearance.
- Self-awareness: Anticipate and avoid OR preemptive measures, for example, wash your hands regularly.

- Computer networks example: Network is congested. Causes: Hard to number them down.
- Sensors nodes are thinner with more resilience and hence the nomenclature is growing rapidly. If we wish to present services through sensors (which we really want to do), we must put the functional and heterogeneity into consideration.

Requirements

The self-management and casual reasoning along with rule-based infrastructure are the enabling technologies of this research.

Self-Healing

As result of the problems mentioned previously, several network management solutions proposed in References 6 through 15 are confined strictly to their respective domains, that is, either mesh network or MANETs. A self-management architecture is proposed in Reference 6 for u-Zone networks. The limited amount of published work currently available for consideration means that there are many unanswered questions. We consider the following questions particularly pertinent:

1. If self-healing is one of the FCAPS (Fault, Configuration, Accounting/Administration, Performance, Security) functions, then what is the physical location of self-healing functions, whether they should function on the gateway or at the client end?
2. How does the control, information, etc. flow from one function to another? Especially, how do self-healing functions interact with the other functions?
3. What are the calling signatures of self-healing functions? If self-healing functions are fault-removing functions, then what are the functions of fault management functions?

4. Are these sub-functions functionally independent? If yes, then there is evidence of a lot of redundancy and if not, then how can self-healing be thought of as an independent entity? In other words, what is the true functionality definition of self-healing?
5. What if the self-management entity itself faces management problems? How should they be tackled?

The autonomic self-management in hybrid networks is a relatively new area of research. In Reference 6, the authors propose an autonomic self-management framework for hybrid networks. Our approach is different from their work in the basic understanding of the functionality of self-management functions. We argue that the self-management functions do not stem from one main set. Rather, they are categorized in such a way that they form on-demand functions and some functions are always-on/pervasive functions.[8] Figure 1.2 gives a clearer description.

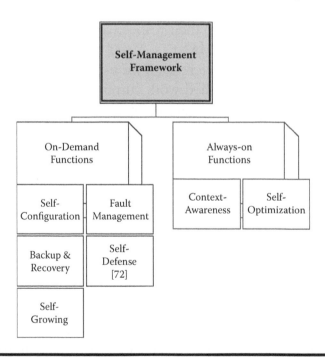

Figure 1.2 The proposed classification of self-management functions.

We propose that context-awareness and self-optimization should be in the always-on category and the others as on-demand functions. This approach is very useful in hybrid environments where there are clients of various batteries and computing powers, memories, etc. and a self-management framework might prove too much middleware. The Normal Functionality Model (NFM) regulates the usage of self-management functions according to computing ability of the client. This gives the client level local self-management. The healing policies come in two parts: condition and action. The condition parts are extracted from the service request that the client sends. The analyzer examines the service requests through the procedures discussed in References 8, 16, 17, 18, and 19.

Causal Reasoning

In order to enable fault identification in self-healing systems, we propose the use of causal reasoning (CR) as an enabling technology. CR was originally proposed in 1976 by Allen Brown at MIT. The CR was used for fault localization intensively in communication systems; that is, radar and sonar detection systems.

In order to understand CR, consider the following example: A doorbell rings. A dog runs through a room. A seated man rises to his feet. A vase falls from a table and breaks. Why did the vase break? To answer this question, one must perceive and infer the causal relationships between the breaking of the vase and the other events. Sometimes, the event most directly causally related to an effect is not immediately apparent (e.g., the dog hit the table), and conscious thought and effort may be required to identify it. People routinely make such efforts because detecting causal connections among events helps them to make sense of the constantly changing flow of events. CR enables people to find meaningful order in events that might otherwise appear random and chaotic, and causal understanding helps people to plan and predict the future.

An important distinction exists between causal perceptions and causal reasoning. Causal perceptions refer to one's ability to sense a causal relationship without conscious and effortful thought. The perceptual information regarding contiguity, precedence, and co-variation underlies the understanding of causality. First, events that are temporally and spatially contiguous are perceived as causally related. Second, the cause precedes the effect. Third, events that regularly co-occur are seen as causally related.

In contrast, CR requires a person to reason through a chain of events to infer the cause of that event. People most often engage in CR when they experience an event that is out of the ordinary. Thus, in some situations, a person may not know the cause of an unusual event and must search for it, and in other situations must evaluate whether one known event was the cause of another. The first situation may present difficulty because the causal event may not be immediately apparent. Philosophers have argued that causal reasoning is based on an assessment of criteria of necessity and sufficiency in these circumstances. A necessary cause is one that must be present for the effect to occur. Event A is necessary for Event B if Event B will not occur without Event A. For example, the vase would not have broken if the dog had not hit the table. A cause is sufficient if its occurrence by itself can bring about the effect (i.e., whenever Event A occurs, Event B always follows). Often, more than one causal factor is present. In the case of multiple necessary causes, a set of causal factors taken together jointly produces an effect. In the case of multiple sufficient causes, multiple factors are present, any one of which by itself is sufficient to produce an effect.

In our approach, we enumerate all the possible conditions and test for the most valid online. The ranking method, which should create mapping between events and solutions, should establish stronger relationships (it is included into the future work). The self-growing nature goes side by side with CR and with bigger size, the performance (precision) of the system increases.

There are many applications of CR. In the area of planning, Wilson[20] uses CR to propose fault tolerant system for the Air Force. Schank[21] proposed his concept of Dynamic Memory as an alternative approach to CR: motivated by the assumption that many (if not most) human activities are governed by applying knowledge from previously encountered situations rather than from abstract rules, Schank proposed Memory-Based Reasoning as a model that was intended both to explain human reasoning and to serve as a basis for computer applications. Subsequently, the term CR was used for a wide range of different approaches to reason and machine learning that sometimes had only one thing in common: the use of past cases for solving present problems. Riesbeck and Schank[22] present a range of small applications that use so-called memory organization packets (MOPs) to reason about cooking recipes, jurisdiction, and the resolution of dispute situations. Kolodner[23] gives a more recent overview of CR that follows the same tradition. The approach presented here can be outlined as follows. A library of past situations is used to solve a present situation. Past situations may have been solved successfully or unsuccessfully. The CR system arranges its library of situations into a network of interlinked memory objects. Links between these memory objects represent abstractions, exemplars, indices, or failures. The system tries to automatically create all these links according to general principles and domain-specific knowledge. Each situation that is presented into the system is compared to those situations already in memory, and is analyzed from the perspective of obtaining a correct solution. This is done by retrieving matching situations solutions from memory and adapting these old situations to new situations. A solution thus obtained can then be tested and its performance analyzed for further refinement. A flowchart of these activities is presented in Reference 24. Retrieval of matching cases can also be used for justification of a solution proposed from outside the system. Kolodner therefore differentiates two types of CR systems: those used for problem solving and those used for understanding problems (Figure 1.3).[23]

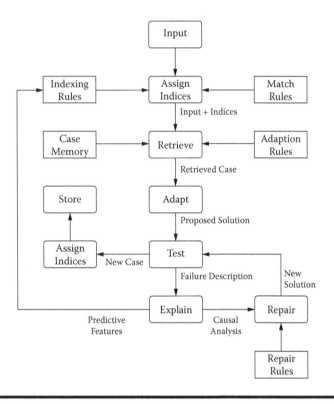

Figure 1.3   A causal reasoning control flow.[23]

We shall discuss the individual steps of CR and compare our technique with it in the following section.

Case Recall

This first step tries to retrieve cases relevant to the case at hand from the case memory. The cases in case memory are indexed by the attributes that are known or were found to be relevant for problem solving. These indices are used to retrieve those cases from memory that share the greatest number of relevant features with the new case, or are very familiar to the new case. Stored cases may be "real" cases or hypothetical cases (e.g., generalizations, composites, etc.) that were added to case memory as a result of previous reasoning. The actual techniques of case retrieval depend on whether a highly hierarchical and interlinked or flat memory structure is chosen by the system designer. Recalling relevant cases can be very time-consuming. In the worst case, all cases in memory have to be matched against the new case. In order to eliminate this possibility, we split the solutions and dynamically compose them into a healing policy. The cases contain the plug-in information and they are matched against the plug-ins through the algorithm proposed in Reference 5. We defined context as the relevant properties of the object. Context Match is a function that matches both service and user contexts and generates the matching factors. There can be many cases but we have shown only three in Figure 1.4. In the first case shown, the User Application Context and Service Application Contexts are

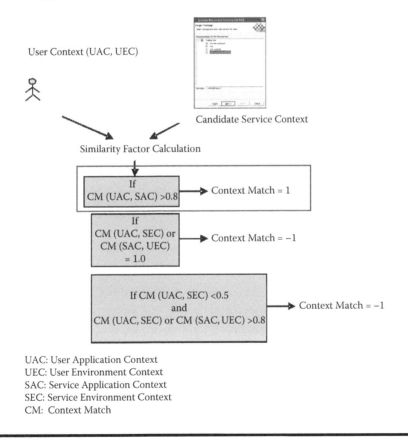

UAC: User Application Context
UEC: User Environment Context
SAC: Service Application Context
SEC: Service Environment Context
CM: Context Match

Figure 1.4   The context orientation process.[5]

matched. If the result is greater than 0.8 (the relevance coefficient which can be changed according to the QoS requirements), it signifies that both resemble each other. In this way, we improve the case recall procedure.

Solution Generation

In this step, the solutions stored with the selected relevant cases are used and adapted to construct a solution for the problem at hand. Since these relevant cases might also have a record of failure stored with them, a warning about potential problems with the new case may also be generated. The way solutions are adapted depends on how solutions are represented, what kind of problems have to be solved, and other criteria. We propose to compose the solutions dynamically.

Testing and Criticizing

In this step, the CBR system tries to assess the usability of the proposed solution before actually applying it to the problem. One method for doing this is to search the case memory for similar solutions to see if they tended to succeed or fail in the past. Another method is to use a simulation of the solution to analyze how the result might deviate from the expected results. This information can then be used to go back and try to repair the solution, or even repeat the previous steps to find a better alternative solution in the case library. In the proposed scheme in this proposal document, we rely on the ranking of the solutions at this stage. However, in the future when the system is mature, we might opt to use the reference-based verification technique, that is to verify the validity through consulting with similar solutions.

Evaluate and Update

This step uses information from results of applying the proposed solution to the real-world problem. If the solution failed (partially), an explanation of why the failure occurred has to be found. Again, the case memory can be searched for cases where a similar failure occurred to adapt the explanation for failure to the current situation. The completed new case can now be added as an additional case to the case memory. Indexing information has to be updated so that the new case will be retrieved in later situations where the knowledge stored with it might be relevant.

Challenges

There are a few challenges that should be addressed while designing a system like the one we have proposed in this research.

Challenge #1: Dynamism and Context-Awareness

The modern-day systems are very dynamic and constantly growing in size. The monitoring of such systems should be done in a way that is useful for reasoning modules that are embedded into the systems. We call this context-awareness in this research document. The response to these changes that are obtained because of this monitoring property is called dynamism.

Challenge #2: Portability

The usefulness of a system depends upon many factors. Portability is one of them. If a system is not portable, it is not considered a general purpose and hence the audience and attention to this system is not wide scale. The consideration of scenarios within one domain is as important as the ones out of domain. While designing such solutions, special attention should be paid to the aspects that it is expandable, extendable, and portable.

Challenge #3: Magnitude of Scale

Ubiquitous systems are large-scale systems. The precision and recall is very important in real-life environment. The speed of matching and response time plays a key role in the backbone of such systems. Therefore, the magnitude of scale is important. Perhaps one idea would be to distribute the intelligence over various nodes and devise a synchronization mechanism for it.

Challenge #4: Heterogeneity

As stated previously, a ubiquitous system would consist of devices having varying capabilities, connected by networks of different characteristics, conforming to different standards, and imposing different requirements on the system. Hence, the situations and conditions related to certain settings are bound to vary. In this situation, a constant self-growing mechanism is required to keep the intelligent repository growing.

Challenge #4: Scale of Uncertainty

The uncertainty in smart systems hampers aggressive fault identification. In a complex distributed network where numerous services are catering to users, the formation of a fault, its identification, and furnishing a solution that fits into the scenario are difficult tasks. With ever-growing scale, the application level faults are primitively fixed through intensive client-server and shaking. We try to eliminate this situation by enforcing casual solutions to the client to fix the application faults.

The requirements presented in this book warrant a new bottom-up system design. The challenges posed by the heterogeneity, longevity, mobility, dynamism, and context-awareness of a ubiquitous system can be handled effectively only at the operating system level. The following areas are (but not limited to) where this research is expected to make significant impact:

1. Hierarchy definition and logical partitioning of the functional components of management functions.
2. Management system for self-healing software.
3. Middleware for thin clients.

4. Service provisioning.
5. Lightweight reasoning.
6. Distributed components integration.
7. Low gateway association.
8. Non-server-side execution.
9. Example of a self-healing software, service request management, and self-healing service integration and execution.

The ubiquity of the system, however, also means that it is difficult to implement and validate a system design without focusing on a concrete example. The dissertation therefore gives a parallel account of the implementation and evaluation of the system at both Multimedia and Computer Architecture Lab in Ajou University and Entity Centric Lab in University of Trento, for u-Frontier project and the OKKAM project as an inspiration to evaluate its performance.

What This Book Is Not

Ubiquitous systems, indeed, are aimed at helping users without being intrusive. Therefore, a large portion of research in ubiquitous systems has focused on human–computer interaction architectures to allow unobtrusive, natural interaction with the system. This book, however, addresses a more fundamental question like how to handle the uncertainty of the real world and most of all how to sort out a pattern that is unusual and find a way out for it.

This book is divided into these chapters:

1. Introduction
2. AHSEN—A Case Study
3. The Proposed Architecture
4. Policy Engine
5. Related Work
6. Implementation Details

7. Prototype
8. Evaluation
9. Conclusion and Future Work
10. Conclusion

References

1. Firetide, www.firetide.com.
2. Burke, R. *Network Management. Concepts and Practice: A Hands-On Approach*, Pearson Education, Inc., 2004.
3. Turing, A. M. On computable numbers, with an application to the Entscheidungs problem, *Proceedings of the London Mathematical Society*, 2(42):230–265, 1936.
4. Doufexi, A., Tameh, E., Nix, A., Armour, S., and Molina, A. Hotspot wireless LANs to enhance the performance of 3G and beyond cellular networks, *Communications Magazine*, 41(7):58–65, 2003.
5. Chaudhry, J. A. and Park, S.-K. Some enabling technologies for ubiquitous systems, *Journal of Computer Science*, 2(8):627–633, 2006.
6. Chaudhry, S. A., Akbar, A. H., Kim, K.-H., Hong, S.-K., and Yoon, W.-S. *HYWINMARC: An Autonomic Management Architecture for Hybrid Wireless Networks*, Network Centric Ubiquitous Systems, 2006.
7. Takahashi, H., Suganuma, T., and Shiratori, N. AMUSE: an agent-based middleware for context-aware ubiquitous services, *ICPADS*, 1:743–749, 2005.
8. Chaudhry, J. A. and Park, S. AHSEN—Autonomic Healing-based Self-management Engine for Network management in hybrid networks, The Second International Conference on Grid and Pervasive Computing (GPC07), 2007.
9. Shehory, O. A self-healing approach to designing and deploying complex, distributed and concurrent software systems. *Lecture Notes in AI*, Vol. 4411, Bordini, R., Dastani, M., Dix, J., and El Fallah Seghrouchni, A., Eds., Springer-Verlag, 2006, pp. 3–11.
10. Trumler, W., Petzold, J., Bagci, F., and Ungerer, T. AMUN: an autonomic middleware for the Smart Doorplate Project, *Personal Ubiquitous Computing*, 10(1):7–11, 2005.

11. Bokareva, T., Bulusu, N., and Jha, S. SASHA: towards a self-healing hybrid sensor network architecture, in Proceedings of the 2nd IEEE International Workshop on Embedded Networked Sensors (EmNetS-II), Sydney, Australia, May 2005.

12. Siewert, S. and Pfeffer, Z. An embedded real-time autonomic architecture, IEEE DenverTechnical Conference, April 2005.

13. Shen, C., Pesch, D., and Irvine, J. A framework for self-management of hybrid wireless networks using autonomic computing principles, in Proceedings of the 3rd Annual Communication Networks and Services Research Conference (Cnsr'05), May 16–18, 2005. CNSR. IEEE Computer Society, Washington, D.C., pp. 261–266.

14. Liu, P. ITDB: an attack self-healing database system prototype, *DISCEX*, 2:131–133, 2003.

15. Zenmyo, T., Yoshida, H., and Kimura, T. A self-healing technique using reusable component-level operation knowledge, *Cluster Computing*, 10(4):385–394, 2007.

16. Garfinkel, S. *PGP: Pretty Good Privacy*, O'Reily & Associates Inc., 1995.

17. Trumler, W., Petzold, J., Bagci, F., and Ungerer, T. AMUN–Autonomic Middleware for Ubiquitous eNvironments Applied to the Smart Doorplate Project, International Conference on Autonomic Computing (ICAC-04), New York, May 17–18, 2004.

18. Gao, J., Kar, G., and Kermani, P. Approaches to building self-healing systems using dependency analysis, Network Operations and Management Symposium, April 19–23, 2004. IEEE/IFIP 1:119–132, 2004.

19. Chaudhry, J. and Park, S. On Seamless Service Delivery, The 2nd International Conference on Natural Computation (ICNC'06) and the 3rd International Conference on Fuzzy Systems and Knowledge Discovery (FSKD'06), 2006.

20. Wilson, K. D. CHEMREG: using case-based reasoning to support health and safety compliance in the chemical industry, *AI Magazine*, 19(1): 1998.

21. Schank, R. C. *Dynamic Memory: A Theory of Reminding and Learning in Computers and People,* Cambridge University Press, 1983.

22. Riesbeck, C. K. and Schank, R. C. *Inside Case-Based Reasoning*, Lawrence Erlbaum Associates, Inc., 1989.

23. Kolodner, J. Reconstructive memory: a computer model, *Cognitive Science*, 7: 4, 1983.

24. Bello-Tomás, J. J., González-Calero, P. A., Díaz-Agudo, B., and Colibri, J. An object-oriented framework for building CBR systems, *ECCBR*, 32–46, 2004.

Chapter 2

Case Study
Autonomic Healing-Based Self-Management Engine

The application area of the scheme that we propose in this book is AHSEN (Autonomic Healing-Based Self-Management Engine). The use of CR can be extremely helpful in fault identification. The implementation of the policy engine and the CR-based fault detection scheme will play an important role in the development and adoption of AHSEN into the ubiquitous networks and self-growing autonomic software systems.

Overview

The vision of ubiquitous computing is classified into two categories: 3C everywhere and physical interaction. 3C consists of computing everywhere, content everywhere, and connectivity everywhere. Physical interaction connects the hidden world of ubiquitous sensors with the real world. With the growing popularity of ubiquitous systems, the importance of enabling technologies and enabling methods is also increasing. The initiative of Open Service Gateway (OSGi) is now confined to the home entertainment and automation industry and newer frontiers of sensor networks have gained the forefront position, for example, Zigbee and 6lowPAN. The extensive research and development in grid computing, especially in the service and convergence areas, has given rise to the idea of connecting sensor networks with grid backbones.

Ubiquitous Networks Industry

The evaluation of different technologies is underway in order to find the best candidate for ubiquitous infrastructure and services. A prominent candidate among the list of probable technologies is hybrid networks, such as u-Zone networks. The ubiquitous-Zone-based networks (u-Zone Networks), on one hand, inherit the features of cluster-based Mobile Ad-hoc NETworks (MANETs), that is, high heterogeneity, mobility, dynamic topologies, limited physical security, limited survivability, and low setup time.[1] On the other hand, they are supported by high-speed mesh backbones.[2] This vast difference in computational ability leaves room for some management framework. The u-Zone networks differ from mesh networks in that they are divided into physical zones and these zones contain clusters within themselves. These clusters are logical interest-based groups that can be strictly centric or distributed within a zone. We needed to divide zones further into clusters because of the following reasons:

1. Relatively smaller size of clusters: We know that ultra-wideband technology-based clusters can be as small as a personal body area network, for example, 6lowPAN. They can be highly mobile and may contain different communication standards than the whole network. In this

situation, it is better to partition them into a separate sub-group in the form of an independent cluster.

2. Seamless connectivity: There can be many dependent devices in a wireless network that are dependent on other devices for their connectivity. If we do not divide the zone into clusters, those devices may have to be connected with the gateway directly which can be a serious setback for their power conservation.

3. Management standards: We intend to apply self-management standards in u-Zone networks. In order to do that, we need to scrutinize the feasibility of the environment. We discovered in our previous research[3] that it would be good to divide zones into clusters.

Open architectures like UC Berkeley's Ninja architecture and Open Service Gateway Initiative (OSGi) are excellent development standards that contain the potential for community-level computing and development. The vastness of domain and complexity of contextual details to document and process and the importance of real-time constraints—time, accuracy, etc—in time- and mission-critical applications that serve as a core of ubiquitous initiative can prove to be capacitive killer applications.

A recent study has revealed that approximately 70% of a system's acquisition cost is related to its management and maintenance. Therefore, an important aspect of these development standards should be to cut down the management cost of systems developed using these standards. The OSGi and Ninja architectures provide administrator-level system management tools that can be automated using scripts and procedures. With applications targeting such a wide scale and hence facing a huge set of exceptions, there is a need for "round-the-clock" administration.

Autonomic computing (AC) presents a novel set of techniques to cut down the management costs in computer systems. The intelligent techniques mimic the behavior of the human mind and reduce the time spent by human administrators in solving the exceptions. The software built to perform autonomic behavior can be called self-managing software. Some examples of self-managing software are:

- Application software such as Tivoli, Lotus, IBM WebSphere, and IBM DB2.
- Hardware-oriented, such as IBM eServer, etc.
- Adaptation layers and sub-modules, such as RoSES, ANUN, AMUSE, HYWINMARC etc.

Dynamic software architectures modify their architecture and enact the modifications during the system's execution. This behavior is most commonly known as run-time evolution or dynamism. Self-managing architectures are a specific type of dynamic software architectures. We define a system that enacts architectural changes at run-time as having a self-managing architecture if the system not only implements the change internally but also initiates, selects, and assesses the change itself without the assistance of an external user. This capability of self-managing software enables self-healing systems to attain a very critical position of solving both the simple and complex problems in the computer systems. IBM has taken initiative in research and development of autonomic systems since 2000.

Autonomic Healing-Based Self-Management Engine (AHSEN)

Software Architecture

In hybrid wireless networks, the network contains high node density, scale, variable topology, and hence mobility issues, and at node level constraints such as high hardware cost, almost all nodes being powered by battery, low processing speed and small memory size, limited transmission range and low bandwidth rates, and scarce power are worth mentioning. This variety of features, constraints, and capabilities poses a greater hindrance to proposing a comprehensive management framework that could accommodate almost all types of nodes. One of the most desired characteristics of such architecture would be that it should be lightweight and could expand its functionality dynamically. This feature is missing in the related literature.

In this section, we present AHSEN. Figure 2.1 shows the client and gateway self-management software architectures. In a gateway-assisted environment, it serves to our advantage to do the bulk of processing, for example, fault analysis, solution composition, etc., at the gateway level and let the client do more important things like self-monitoring, optimization, and local self-management using normal functionality model (NFM).

When a mobile device comes in the range of a certain gateway, upon configuration request it is provided with an NFM, which is specially designed for certain device classes. The NFM runs a self-check and reports to the gateway self-management framework (SMF) that issues a client version of

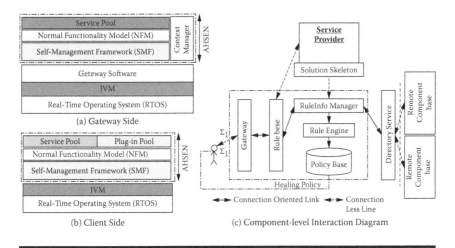

Figure 2.1   The AHSEN architectures of the gateway (a), the client (b), and component-level interaction diagram (c).

AHSEN. The client AHSEN consists of two directory services for services and plug-ins, client SMF responsible mainly to host services and plug-ins, and NFM.

The plug-in and service pool contain the directory information of related services provided by different service providers. At the client side, the plug-in manager hosts the plug-ins downloaded from remote locations. At the server side, it contains the directory service that contains the plug-in information. The NFM is a device-dependent ontology that is downloaded, along with SMF, on the device at network configuration level. It provides a mobile user with an initial default profile at the gateway level and device level functionality control at the user level. The NFM contains specifications of normal range of functional parameters of the device, services environment, security certificates, and network standards. The SMF constantly traps the user activities, sends them to the SMF at the gateway, and hosts the executables. The SMF at the gateway directs the trap requests to the context manager who updates the related profile of the user.

Besides the periodic context update (Σ_i), the client SMF also sends anomaly reports to server SMF (Σ_j), which is forwarded to the rule-base for case-based reasoning.[4] If the match is found, its related healing policy is forwarded to the mobile device. In case the client SMF reports the problem, heuristics are used to determine the cause and solution for the problem reported. The Ruleinfo Manager collects the results from the rule-base, solution skeleton (which is a policy guideline provided by a third-party vendor for problem resolution) and matches it against the components information listed in the directory service using the scheme proposed in Reference 5. The rule engine[6] composes the components together and generates a healing policy, which is stored in the policy base and eventually forwarded to the mobile device. This way, the self-managing software grows its knowledge repository. The benefit of using the rule engine is that for every condition there can be several execution components. When the condition part is prepared in the rule engine, a part of it is sent to the rule base. By doing this we increase the rule repository and increase the probability of a fault being detected. The rest of the policy is stored in the policy repository for a mobile device's direct use and the condition part serves an extra job of fault detection in rule-base.

A detailed architecture of the SMF is shown in Figure 2.2.

The root cause analyzer (RCA) is the core component of the problem-detection phase of healing. The state transition analysis based approaches[7] might not be appropriate as hidden Markov models (HMMs) take long training time along with exhaustive system resources utilization. The profile-based root cause detection might not be appropriate mainly because of the vast domain of errors expected. Considering this situation, we use the meta-data obtained from NFM[8] to trigger finite state automata (FSA) series present at the RCA. In the future, we plan to modify the state transition analysis tool[7] along the lines of fault analysis domains. After analyzing the root-cause results from the RCA, the root cause fragmentation (RCF) manager in cooperation with the signature repository and scheduler search for the already developed solutions; otherwise, it arranges a time slot based scheduler for plug-ins. The RCF manager uses heuristics and gathers all the possible problem

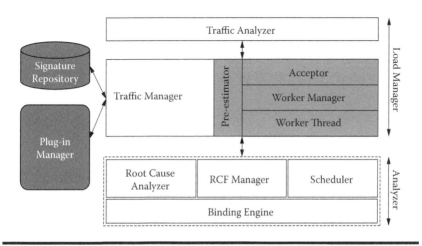

Figure 2.2   Architecture of self-management framework (SMF).

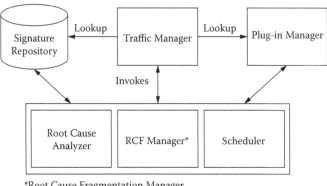

Figure 2.3   A self-management framework core.

causes and their solutions from the already given rule base and hands the context to the scheduler. The scheduler uses the time slot based mechanism[9] to gather component context and forward it to a binding engine. The binding engine generates an XML file containing the execution sequence and feedback mechanism. We call this part an analyzer and use it for autonomic self-management in ubiquitous systems. The traffic manager manages the traffic directed to the service gateway and consists of sub-parts as shown in Figure 2.3.

AHSEN Components

The detailed description of some components of AHSEN is as follows.

Normal Functionality Model (NFM)

The NFM is a policy file for the clients consisting of three components.

1. Device Normal Functionality Model: It contains all the normal exceptions for that device; that is, this device may

face this type of error while using certain types of services and this is the solution. Initially, the device NFM
would be empty but as the gateway repository grows
(because of constant monitoring of client activities), its size
would grow as the exceptions for a device would grow.

2. Network Normal Functionality Model: This part contains
the hash keys of all the access points (AP) of that zone
(the whole area is divided into zones and zones are physical buildings so we can say that Network NFM contains
the information about all the APs of the building).

3. Common Business Policies: It is a kind of "switchboard"
or "main form" that contains the list of applications/ser-
vices offered for clients in that zone.

Quality of Service (QoS) Map: We have some rules that if
the device contains this type of hardware configuration (i.e.,
CPU, memory, etc.) it will have a certain QoS matrix. The
values of that matrix would be used to measure if the device
were working within its optimal working capacity or below
that throughput expectations. The matrix will contain the
network bandwidth, link speed, etc. and some client dependent variables, that is, optimal CPU usage range (80% to 95%),
memory allocation (30% free), optimal number of services
concurrently running on that device (dependent on client
resources).

Network Map: This part contains the hash keys of all the
access points that are associated with the host gateway. This
way a mobile user can move freely within a zone without having to worry about manual configuration and having to have
seamless configuration and mobility alongside exclusive access
to a network.

Service Map: It contains a list of available series that will be
used to display those to the user. This data can be accessed
from the service registry inventory.

Self-Management Framework (SMF)

The SMF consists of a traffic manager that redirects the traffic to all parts of the SMF. As proposed in Reference 10, the faults can be single root-cause based or multiple root-cause based. We consider this scenario and classify an RCA that checks the root cause of failure through the algorithms proposed in Reference 11. After identifying the root causes, the RCF manager looks up the candidate plug-ins as a solution. The RCF manager also delegates the candidate plug-ins as possible replacement of the most appropriate. The scheduler schedules the service delivery mechanism as proposed in Reference 12. The processed fault signatures are stored in a signature repository for future utilization. The plug-in manager is a directory service for maintaining the latest plug-in context. This directory service is not present at the client level. In Reference 2, the authors classify self-management into individual functions and react to the anomaly detected through SNMP messages. The clear demarcation of self-* functions is absent in modern-day systems as there is no taxonomy done for various fault types. This is one of the main reasons why we prefer component integration to conventional high granularity modules for self-management.[5] A detailed architecture of the SMF is shown in Figure 2.3.

The RCA plays a central part in the problem detection phase of self-healing. The state transition analysis based approaches[7] might not be appropriate as HMMs take a long training time along with their "exhaustive" system resources utilization. The profile-based root cause detection might not be appropriate mainly because the domain of errors expected is very wide.[3,13,14] We use the metadata obtained from NFM to trigger a finite state automata (FSA) series present at RCA. In the future, we plan to modify the state transition analysis tool[7] according to fault analysis domains.[15] After analyzing the root cause results from the RCA, the RCF manager, the signature repository, and the scheduler search the already developed solutions for a particular fault. If not, it arranges a time slot based scheduler as proposed in Reference 16 for plug-ins. The traffic manager directs the traffic toward different parts of AHSEN.

Service Lifecycle

Service Registration: When a service provider makes a service, it aims to make that service available to the users. The provider makes it available to the users online. To let users know about their services, they register those services to service gateways. The service provider provides the location information of the service and some metadata. The metadata can be in the form of delivery certificates (the certificates and private keys needed to access service), key words, etc.

Service Representation: Every service contains some time slot reservation mechanism for service consistency. This process includes a sub-process called binding. An example of this procedure follows. Binding is the process of reservation of service version at the gateway. This is a period of time in which the service is reserved and no new version of the service can be launched. This time, for service, is called time to live (TTL). In Table 2.1, different source services have various services preceding which creates a string of services. The TTL is calculated

Table 2.1 The Binding Table

SID	TSID	Parameters	TTL
C001455D	C001456D	$t_1, s_1, w[w_1, w_{n-1}]$	001589GHJ+3000
C001456D	C001457D	$T_1, T_2, s_2, w[w_1, w_{n-1}]$	001589GHJ+3000
C001457D	C001458D	$T_2, T_3, s_3, w[w_1, w_{n-1}]$	001589GHJ+3000
C001458D	C001459D	$T_3, T_4, s_4, w[w_1, w_{n-1}]$	001589GHJ+3000
C001459D	C001455D	$T_4, T_5, s_5, w[w_1, w_{n-1}]$	001589GHJ+3000
C001455D	C001456D	$T_5, T_6, s_6, w[w_1, w_{n-1}]$	001590GHJ+3000

Note: SID = Service ID; TSID = Target Service ID; w = weights as par.

from the binding time and 3000 ms. Calculation of TTL reserves one service and the service clashes are eliminated through it.

In mPRM, the services can be accessed from a remote location. That is why we have a virtual service pool (VSP) hosting the services that are used from a remote location. The VSP is activated and an LDAP search is performed when no suitable service is found at the local service pool (LSP).

Service Composition: The services can be composed, which means that the user can open a service composer, which is a graphical interface, and connect services with each other. Services might have compatibility problems, for example, one service is in-taking three parameters, but the service connecting to it is out-putting two parameters. The user has to deal with these issues and provide the missing parameters or deduct the additional parameters. The composer's job is to notify the user about it and make sure that consistency is enforced.

Service Delivery: Now the services are presented to the user in a tree manner; in other words, the main form (switchboard) will appear to the user from where the user can choose to go and select the services of his or her own choice. An example follows.

Building A is a zone with $[A_1, A_2 \ldots A_n]$ as various enterprises. The main board will contain listings of $[A_1, A_2 \ldots A_n]$ and the user can choose to browse through A_1 or A_n, etc. Once the user selects the required service, the gateway "mediates" between the service provider and the user in establishing the connection. Once the connection is established, the user will access the service provided from the service provider and keep listening to the activities at the client end.

Application Scenario

The traffic manager receives SOAP requests from many devices within a cluster and redirects them to all the other internal parts of SMF. Figure 2.4 shows the structure of delay time-based peak load control. The acceptor thread of the traffic manager receives a SOAP request (service request)

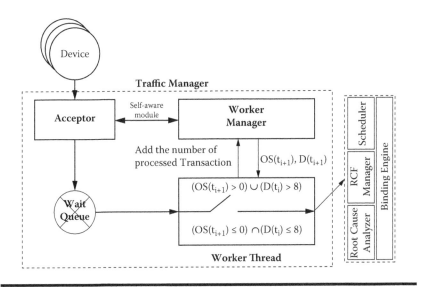

Figure 2.4   Structure of delay-time based peak load control.

and then puts it into the wait queue. The wait queue contains the latest context of the gateway load. If the gateway
is in saturated state, the self-aware sub-module handles the
service request. Figure 2.5 is the pseudo code of self-aware
sub-module in WorkerManager's Delay Time Algorithm. Let a
service request (SR1) arrive at the gateway. At first, the SR1 is
checked to see if it contains the comebacktime stamp (for fair
scheduling). If the comebacktime is "fair," that is, the service
request is returned after the instructed time, it is forwarded
to the wait queue. Otherwise, it is accessed against the workload of the worker manager. The acceptor is updated about
the latest status of the worker manager. The acceptor evaluates the intensity of current workload (how long it will take
to free resources) and adds buffer (buffer is the time to give
some extra room to gateway) to the time. The aggregate time
is assigned to SR1 and the service request is discarded. When
the system is ready to accept the service request, the traffic
manager gets a worker thread from a thread pool and runs
it. The worker thread gets the delay time and the over speed
from the WorkerManager. The admission to other internal

parts SMF is controlled by the worker thread that accepts the arriving requests only if the over speed $OS(t_{i+1})$ at the time t_{i+1} is below zero and the delay time $D(t_i)$ at the time t_i is below the baseline delay δ. Otherwise, the requests have to sleep for the delay time calculated by the WorkerManager. After the worker thread sleeps for the delay time, the worker thread redirects the requests to the RCA, the RCF manager, and the scheduler. Finally, the worker thread adds the number of processed transactions after finishing the related transaction. After sleeping during interval time, the WorkerManager gets the number of transactions processed by all worker threads and the maximum transaction processing speed configured by a system administrator. Then, the WorkerManager calculates the transactions per milliseconds (TPMS) by dividing the number of transactions by the maximum transaction processing speed and calculating the over speed $OS(t_{i+1})$, which means the difference of performance throughput at the time t_{i+1} between the TPMS and the maximum transaction processing speed during the configured interval time. If the value of the over speed is greater than zero, the system is considered to be in an overload state. Accordingly, it is necessary to control the overload state. On the contrary, if the value of the over speed is zero or less than zero, it is not necessary to control the transaction processing speed. For controlling the overload state, use the delay time algorithm of the WorkerManager. Figure 2.5 describes the formulas for calculating the delay time.

If the over speed $OS(t_{i+1})$ is greater than zero, the first formula of Figure 2.5 is used for getting a new delay time $D(t_{i+1})$ at the time t_{i+1}. The $N(t_{i+1})$ means the number of active worker threads at time t_{i+1} and $D(t_i)$ means the delay time at time t_i. If $D(t_i)$ is zero, $D(t_i)$ must be set to one. If $OS(t_{i+1})$ is below zero and the delay time $D(t_i)$ at time t_i is greater than the baseline delay δ, $D(t_{i+1})$ is calculated by applying the second formula of Figure 2.5. On the contrary, if $D(t_i)$ is below the baseline delay, $D(t_{i+1})$ is directly set to zero. In other words, because the state of system is under load, the delay time at time t_{i+1} is not

Let a service request SR_1 arrives at Acceptor
Check SR$_1$.comebacktime
If (*SR$_1$.comebacktime*='fair') // check the virtual queue
        Wait Queue ← Send *SR$_1$*
Acceptor ← Send *current_context(worker_Manager_Status_Update)*
If (*current_context*!='overloaded') Wait queue ← Send *SR$_1$*
else
        While (*current_context*='overlaoded')
                delaytime= Calculate (intensity_of(*current_context*)+*buff*
                Set *SR$_1$.comebacktime* ← *delaytime*, Dismount *SR$_1$*
While run_flag equals "true" do
get interval time for checking load, sleep during the interval time
get the number of transactions processed during the interval time
get the configured maximum speed, TPMS := number of transactions / interval time
over speed := TPMS − the configured maximum speed
If over speed >0 then, get the previous delay time
        if previous delay time = 0, previous delay time := 1
        get the number of active worker thread, new delay time:= over speed / number
            of active worker * previous delay time
else
        get current delay, if current delay > δ, new delay time := current delay *
        else

Figure 2.5 **The pseudo code for self-aware module and workerman-ager's delay time algorithm.**

necessary. Accordingly, the worker thread can have admission to other internal parts of SMF. The baseline delay is used for preventing repetitive generation of the over speed generated by suddenly dropping the next delay time in a previous heavy load state. When the system state is continuously in a state of heavy load for a short period of time, it tends to regenerate the over speed to suddenly increment the delay time at time t_i and then suddenly decrement the delay time zero at time t_{i+1}. In other words, the baseline delay decides whether the next delay time is directly set to zero.

The β percent of the second formula of Figure 2.6 decides the slope of a downward curve. However, if the delay time at time t_i is lower than the baseline delay, the new delay time at time t_{i+1} is set to zero. Accordingly, when a system state becomes the heavy overload at time t_i, the gradual decrement

$$D(t_{i+1}) := \begin{cases} \dfrac{OS(t_{i+1})*D(t_i)}{N(t_{i+1})}, \textit{if}\left(OS(t_{i+1})\right)>0 \\[2em] D(t_i)*\beta, \textit{if}\left(\left(OS(t_{i+1})\right)\leq 0\cap\left(D(t_i)>\delta\right)\right), \\[2em] 0, \textit{if}\left(\left(OS(t_{i+1})\right)\leq 0\cap\left(D(t_i)\leq\delta\right)\right) \end{cases}$$

Figure 2.6   A mathematical model for delay time calculation.

by β percent prevents the generation of repetitive over speed caused by abrupt decrement of the next delay time.

Once the service request is received by the worker thread, the analysis of the cause of anomaly starts. As proposed in Reference 10, the faults can be single root cause based or multiple root cause based. We consider this scenario and classify an RCA that checks the root failure cause through the algorithms proposed in Reference 11. After identifying the root causes, the RCF manager looks for the candidate plug-ins as a solution. The RFC manager also delegates the candidate plug-ins as possible replacements of the most appropriate plug-in software that are available in the plug-in bank. The scheduler schedules the service delivery mechanism as proposed in Reference 12. The processed fault signatures are stored in a signature repository for future utilization. Let N be concurrent service requests at the server at full time. This means that as soon as one thread finishes its execution, a new one will take its place. This assumption is made to ensure that we analyze the worst-case scenario of performance time with N service requests in the execution queue. This means that the execution of a service request is done from its first until its last quantum (subparts of a service request, i.e., analyze, evaluate, categorize, etc.) in the presence of other $N–1$ service requests.

- K * index variable spanning the service requests: $1 \le k \le N$
- S_j * CPU quantum length for server j ($S_{isolated}$ represents that value for a specific isolated server)
- I_{kj} * the number of cycles the kth service request needs to complete on server j
- $i_{k\_isolated}$ represents that value for a specific isolated server

When a transaction is completed, and considering that the kth client permanently issues the same request to the server, then the number of transactions (T_x) that may be completed for k clients in interval T is

$$Tx_k = \Delta * \frac{T}{t_{k\_isolated} * N} \text{ or } Txcpu_k = \Delta * \frac{T}{t_{k\_isolated} * N}$$

in terms of CPU utilization. It means that the execution time for a transaction depends upon the number of service requests in an active queue. Therefore, we can calculate the estimated time a CPU needs in order to get free from the requests in an active queue.

Let there be N number of service requests present in the active queue. In time t_{k1}, the service request s_{k1} is being executed, the K_{N-1} service requests will reside in the memory.

- Ms_k * memory size kth service request. $Ms_k \ge 1$
- B_k * branch statements in Ms_k in kth service request. $B_k \ge 0$
- $Ttpb_k$ * the time needed per transaction

$$Txmem_k = \varphi * \left(\frac{Ms_k * B_k}{Ttpb_k} * N \right)$$

where φ is coefficient of CPU utilization.

Let u(t) denote load of service request. We can normalize the service request as

$$y(t) = \frac{u(t) - u_{min}(t)}{u_{max}(t)} u_{min}(t) u_{max}(t) T_k T_k k = 0,1,2,\ldots,K, y_{T_k}(t)$$

where and denotes the minimum and maximum load of the service request.

Different service request traces can be compared with each other, while the impact of their internal analysis is eliminated. If we define T_k as the kth threshold for, then a function is defined by

$$y_{T_k}(t) = \begin{cases} 1 & y(t) \geq T_k \\ 0 & else \end{cases}$$

$$T_k = k \frac{1}{K} y_{T_k}(t) = 1$$

$$t_k \leq \frac{u(t) - u_{min}(t)}{u_{max}(t)}$$

or,

$$T_{nwbuff_k} = \delta * \left\{ \frac{u(t) - u_{min}(t)}{u_{max}(t)} \right\}$$

where δ is coefficient of CPU utilization.

$$T_{comeback} = Txcpu_k + Txmem_k + T_{newbuff_k}$$

$$T_{comeback} = \left\{ \Delta * \frac{\tau}{t_{k\_isolated} * N} \right\} + \left\{ \varnothing * \left(\frac{Ms_k * B_k}{Ttpb_k} * N \right) \right\} +$$

$$\left\{ \delta * \left(\frac{u(t) - u_{min}(t)}{u_{max}(t)} \right) \right\}$$

Now we know that the service request has exponential distribution and the arrival rate has Poisson distribution. Therefore, we can say that the round trip delay (RTD)

$$RTD = T_{\text{comeback}} + queuing\ delay + \tau_f + \tau_e$$

where τ_f is the propagation delay from server to the bottleneck link buffer that is the gateway buffer, and τ_e is the propagation delay over the return path from the bottleneck link buffer to the client (the service request generator).

$$RTD = \left\{ \Delta * \frac{T}{t_{k_{isolated}} * N} \right\} +$$

$$\left\{ \varnothing * \left(\frac{Ms_k * B_k}{Ttpb_k} * N \right) \right\} +$$

$$\left\{ \delta * \left(\frac{u(t) - u_{min}(t)}{u_{max}(t)} \right) \right\} + queuing\ delay +$$

$$\tau_f + \tau_e \left\{ \delta * \left(\frac{u(t) - u_{min}(t)}{u_{max}(t)} \right) \right\} + queuing\ delay + \tau_f + \tau_e$$

Efficiency is calculated as

$$E(S * N) = \frac{N * T_s}{S * 1 * T(S * 1)} \approx \frac{N * T_s}{(S)S * T}$$

In this expression, the range for S is

$$\left[1 .. \frac{N}{n} \right]$$

and the range for N is [1..k*$\log_n N$], which is less than [2$\log_n N$]' for HYWINMARC-based,[2] and

$$\left[2\log_n \frac{N}{2} + N^{n^2} \right]''$$

for RoSeS-based[17] solutions.

Simulation Results

In order to prove performance stability of the self-aware PLC-based autonomic self-healing system, we simulated the self-aware delay time algorithm of the WorkerManager. As for load generation, the LoadRunner 8.0 tool is employed. The delay time and over speed are used as a metric for simulation analysis. The maximum speeds, δ and β, for delay time algorithm are configured 388, 100 ms, and 0.75 respectively. Figure 2.7 shows the result of simulation for describing the relationship between the over speed and the delay time after the saturation point.

These experimental results prove that the proposed delay time algorithm of the WorkerManager has an effect on controlling the over speed. As the number of concurrent users is more than 220 users, over speed frequently takes place. Whenever over speed happens, each worker thread sleeps for the delay time calculated by the WorkerManager. As the higher over speed takes place, each worker thread sleeps for more time so that the over speed steeply goes down. Although the over speed steeply goes down, the delay time does not steeply

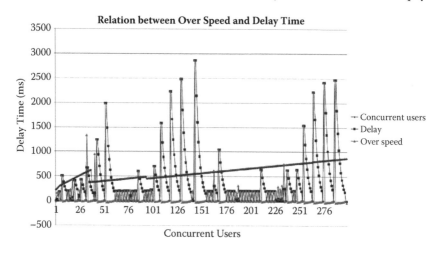

Figure 2.7   Simulation results showing the effect of delay time algorithm.

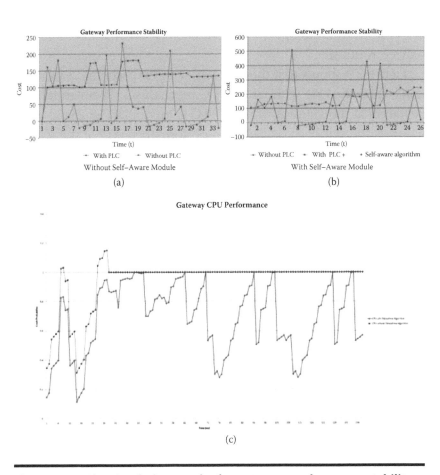

Figure 2.8   The simulation results for gateway performance stability.

go down due to the baseline delay value, δ. As the baseline delay value is set to 100 ms in this experiment, the delay time gradually goes down until the 100 ms. As soon as the delay time passes 100 ms, the next delay time is directly set to zero. The result of the simulation in Figure 2.8 shows that the over speed does not happen until zero delay time due to the slope of a downward curve. However, as soon as the delay time passes zero, the over speed again occurs and the next delay time controls the over speed.

Although the heavy request congestion happens in a traffic manager of the gateway, the delay time based PLC mechanism can prevent the thrashing state in overload phase and help the traffic manager to execute stably the management service requests.

Figure 2.9 shows that the gateway with PLC scheme is more stable than the one without the PLC mechanism. The standard deviation at the gateway without PLC is more than 58.23, whereas the standard deviation in performance cost at the gateway with the PLC mechanism is 24.02. This proves the argument posted in the previous section that the PLC mechanism provides stability to gateways in u-Zone-based networks. Figure 2.9 shows that applying a self-aware sub-module to the PLC mechanism gives a stable performance than applying a PLC algorithm only. The stability in the cost function with time shows that the cost is predictable over a time scale. Although the cost of applying a PLC mechanism with self-aware module is more than without it, the self-aware PLC gives more stability and hence is more suitable in unpredictable, dynamic, and highly heterogeneous u-Zone networks. Figure 2.8(a) shows the gateway CPU performance. Whenever the gateway probability rises up, it is reduced by the algorithm proposed here whereas the gateway without the self-aware algorithm goes into a crash state.

Figure 2.9 shows the results obtained from different experiments with varying parameters. We observe that increase in the users increases the throughput. Increasing the β increases the capacity of the gateway to entertain more dense flux of service requests and hence increases the throughput. According to our experiments, the most important factor is the maximum speed configuration of the WorkerManager, while α and β are not directly related to the efficiency of the WorkerManager. Figure 2.8(c) gives the illustration of CPU usage and estimation of throughput when the proposed algorithm is in use and when it is not in use. We can observe that when the maximum speed of WorkerManager is low, the throughput is low and vice versa.

Figure 2.10 shows the comparison of resource utilization with the throughput of the system. In this experiment, our test bed consists of two sensor nodes, one laptop as mobile nodes,

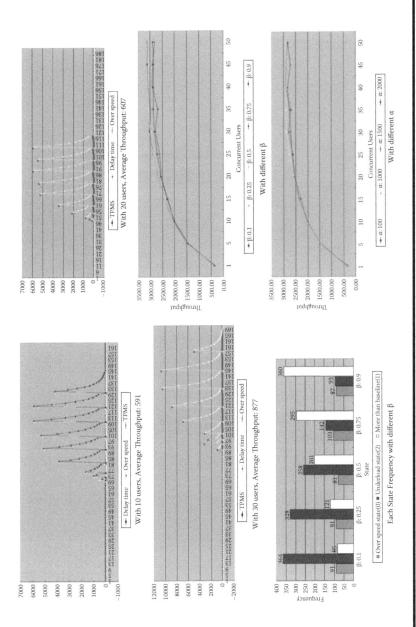

Figure 2.9   The simulation results for gateway performance stability using workermanager delay time algorithm.

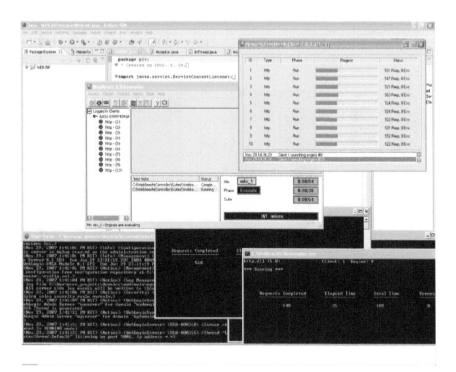

Figure 2.10   The test bed for autonomic self-aware service request optimization scheme.

and a desktop computer as a server equipped with BEA web logic platform, JSP-based server pages, workbench 5.0 client and controller, and media streaming applications. The results show that there is a marked difference between the maximum throughput with no sleep times and throughput with different parameters and in the presence of sleep times.

References

1. Doufexi, A., Tameh, E., Nix, A., Armour, S., and Molina, A. Hotspot wireless LANs to enhance the performance of 3G and beyond cellular networks, *Communications Magazine*, 41(7): 58–65, 2003.

2. Chaudhry, S. A., Akbar, A. H., Kim, K.-H., Hong, S.-K., and Yoon, W.-S. *HYWINMARC: An Autonomic Management Architecture for Hybrid Wireless Networks*, Network Centric Ubiquitous Systems, 2006.

3. Lunt, T. F. Real-time intrusion detection, in Proc. COMPCON, San Francisco, CA, Feb. 1989.

4. Anglano, C. and Montani, S. Achieving self-healing in service delivery software systems by means of case-based reasoning. *Applied Intelligence*, To appear.

5. Ma, J., Zhao, Q., Chaudhary, V., Cheng, J., Yang, L. T., Huang, H., and Jin, Q. *Ubisafe Computing: Vision and Challenges* (I), Springer LNCS Vol. 4158, Proc. of ATC-06, 2006.

6. Lee, Y., Chaudhry, J. A., Min, D., Han, S., and Park, S. A dynamically adjustable policy engine for agile business computing environments, *Lecture Notes in Computer Science, Advances in Data and Web Management*, Joint 9th Asia-Pacific Web Conference (APWeb/WAIM 2007), 785–796.

7. Ilgun, K., Kemmerer, R. A., and Porras, P.A. State transition analysis: a rule-based intrusion detection approach, *IEEE Transactions on Software Engineering*, 21(3):181–199, 1995.

8. Garfinkel, S. *PGP: Pretty Good Privacy*, O'Reily & Associates Inc., 1995.

9. Chaudhry, J. A. and Park, S.-K. Some enabling technologies for ubiquitous systems, *Journal of Computer Science*, 2(8):627–633, 2006.

10. Trumler, W., Petzold, J., Bagci, F., and Ungerer, T. AMUN–Autonomic Middleware for Ubiquitious eNvironments Applied to the Smart Doorplate Project, International Conference on Autonomic Computing (ICAC-04), New York, May 17–18, 2004.

11. Gao, J., Kar, G., and Kermani, P. Approaches to building self-healing systems using dependency analysis, Network Operations and Management Symposium, April 19–23, 2004. IEEE/IFIP 1:119–132, 2004.

12. Chaudhry, J. and Park, S. On Seamless Service Delivery, The 2nd International Conference on Natural Computation (ICNC'06) and the 3rd International Conference on Fuzzy Systems and Knowledge Discovery (FSKD'06), 2006.

13. Lunt, T. F. et al. A real-time intrusion detection expert system, SRI CSL Tech. Rep. SRI-CSL-90-05, June 1990.

14. Lunt, T. F. et al. A real-time intrusion detection expert system (IDES), Final Tech. Rep., Comput. Sci. Laboratory, SRI Int., Menlo Park, CA, Feb. 1992.

15. Radosavac, S., Seamon, K., and Baras, J.S. Short paper: buf-STAT—a tool for early detection and classification of buffer overflow attacks, security and privacy for emerging areas in communications networks, First International Conference on SecureComm, Sept. 5–9, 2005, pp. 231–233.

16. Hu, J., Pyarali, I., and Schmidt, D. C. Applying the proactor pattern to high-performance web servers, in Proceedings of the 10th International Conference on Parallel and Distributed Computing and Systems, IASTED, Oct. 1998.

17. Shelton, C. and Koopman, P. Improving system dependability with alternative functionality, DSN04, June 2004.

Chapter 3

The Proposed Architecture

Introduction

The self-healing system cannot function without the support of intelligent components. There are numerous approaches proposed in the literature, discussed in earlier chapters, to improve the reasoning facility for intelligent systems. The biggest hurdle in real-time user interactive systems is to discriminate between normal and abnormal behavior. We propose a mechanism that could help in discriminating normal behavior from abnormal behavior. Figure 3.1 gives an overview of our approach.

Figure 3.1 shows that we store the functional specifications of software under study. We can use various software CASE tools to export the functional specification from design document into XML format. These XML files stay consistent throughout the process specified in this document. The reason for storing the specifications into XML format is that once in XML format, the specifications can be used for many

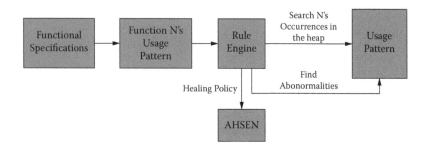

Figure 3.1   A process model.

development purposes, inter-system communication standard, and XML is a business software standard (i.e., BPML, which is derived from XML).

Process Conversion to XML Documents

In a software development life cycle, the design document is the document that gives the visual specifications about the processes, roles, and goals to achieve. Using case tools, we can generate their XML description. We have used ADONIS® software to generate XML files from the business specifications. Consider Figure 3.2 as an example of a process.

In Figure 3.2, a set of processes is shown. There are classes and functions that are made by programmers and every programmer has a team lead. The classes can be hierarchical and the programmers can be called coders. Using CASE tools, we generate an XML description of the model as shown in Figure 3.3.

The exported XML form of the process is shown in Figure 3.4.

Now during the life cycle of the software, these XML files would be considered as the base standard of operations. A tree can be generated using JDOM libraries. It is also used as a library in the experiment performed to back this research. JDOM is an open source, tree-based, pure Java API

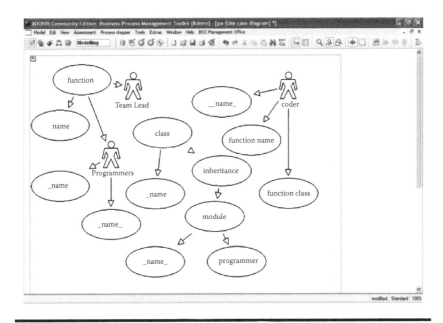

Figure 3.2   Screen shot: A sample model for process specification extraction.

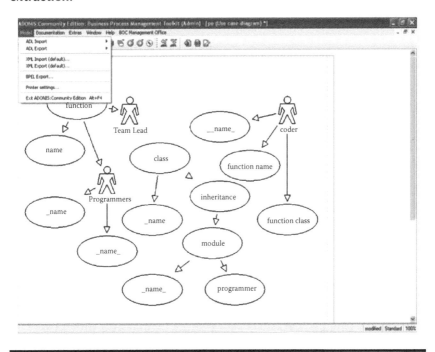

Figure 3.3   Specification export process.

```
<?xml version="1.0" encoding="UTF-8" ?>
<!DOCTYPE ADOXML (View Source for full doctype..)>
<ADOXML version="3.1" date="25.09.2008" time="16:47" database="adonisdb" username="Admin" adoversion="Version 3.9">
- <MODELS>
  - <MODEL id="mod.13204" name="Programmer Code" version="0.0" modeltype="Business process model" libtype="bp"
    applib="ADONIS:CE BPMS BP Library">
    - <MODELATTRIBUTES>
      <ATTRIBUTE name="Version number" type="STRING">0.0</ATTRIBUTE>
      <ATTRIBUTE name="Author" type="STRING">Admin</ATTRIBUTE>
      <ATTRIBUTE name="Creation date" type="STRING">25.09.2008, 16:13</ATTRIBUTE>
      <ATTRIBUTE name="Date last changed" type="STRING"/>
      <ATTRIBUTE name="Last user" type="STRING"/>
      <ATTRIBUTE name="Keywords" type="STRING"/>
      <ATTRIBUTE name="Comment" type="STRING"/>
      <ATTRIBUTE name="Model type" type="ENUMERATION">Current model</ATTRIBUTE>
      <ATTRIBUTE name="State" type="ENUMERATION">In process</ATTRIBUTE>
      <ATTRIBUTE name="Reviewed on" type="STRING"/>
      <ATTRIBUTE name="Reviewed by" type="STRING"/>
      <ATTRIBUTE name="Description" type="STRING"/>
      <ATTRIBUTE name="Number of objects and relations" type="INTEGER">0</ATTRIBUTE>
      <ATTRIBUTE name="World area" type="STRING"/>
      <ATTRIBUTE name="Grid" type="STRING"/>
      <ATTRIBUTE name="Zoom" type="INTEGER">0</ATTRIBUTE>
      <ATTRIBUTE name="Viewable area" type="STRING"/>
      <ATTRIBUTE name="Current mode" type="STRING"/>
      <ATTRIBUTE name="Base name" type="STRING">Programmer Code</ATTRIBUTE>
      <ATTRIBUTE name="Access state" type="ENUMERATION">write</ATTRIBUTE>
      <ATTRIBUTE name="Current page layout" type="STRING"/>
      <ATTRIBUTE name="Connector marks" type="STRING"/>
      <ATTRIBUTE name="Type" type="STRING">Business process model</ATTRIBUTE>
      <ATTRIBUTE name="Change counter" type="INTEGER">7</ATTRIBUTE>
      <ATTRIBUTE name="Font size" type="INTEGER">100</ATTRIBUTE>
      <ATTRIBUTE name="Context of version" type="STRING">0.0</ATTRIBUTE>
      <ATTRIBUTE name="Reference overview active" type="INTEGER">0</ATTRIBUTE>
      <ATTRIBUTE name="Products" type="INTEGER">0</ATTRIBUTE>
      <ATTRIBUTE name="RACI/DEMI visualisation" type="INTEGER">0</ATTRIBUTE>
      <ATTRIBUTE name="Responsible role" type="INTEGER">0</ATTRIBUTE>
      <ATTRIBUTE name="Input/Output" type="INTEGER">0</ATTRIBUTE>
      <ATTRIBUTE name="Use cases" type="INTEGER">0</ATTRIBUTE>
      <ATTRIBUTE name="IT systems" type="INTEGER">0</ATTRIBUTE>
      <ATTRIBUTE name="Modelling direction" type="ENUMERATION">horizontal</ATTRIBUTE>
      <ATTRIBUTE name="Products color" type="STRING">whitesmoke</ATTRIBUTE>
      <ATTRIBUTE name="RACI/DEMI color" type="STRING">whitesmoke</ATTRIBUTE>
      <ATTRIBUTE name="Responsible role color" type="STRING">whitesmoke</ATTRIBUTE>
      <ATTRIBUTE name="Input/Output color" type="STRING">whitesmoke</ATTRIBUTE>
      <ATTRIBUTE name="Use cases color" type="STRING">whitesmoke</ATTRIBUTE>
      <ATTRIBUTE name="IT systems color" type="STRING">whitesmoke</ATTRIBUTE>
      <ATTRIBUTE name="Contact person" type="STRING"/>
      <RECORD name="Change history"/>
    </MODELATTRIBUTES>
  </MODEL>
</MODELS>
```

Figure 3.4   An extracted XML specification model.

for parsing, creating, manipulating, and serializing XML documents. It is more complete than either SAX (which does not
offer any standard way to write new XML documents) or DOM
(which can manipulate XML documents but does not know
how to parse or serialize them). It is also much easier to use
than either SAX or DOM for most tasks. It has the convenience
of a pull-based tree API with DOM and the familiarity of following standard Java conventions with SAX. However, JDOM
is not SAX and it is not DOM. JDOM can transfer data to and
from SAX and DOM, but it is its own API, complete unto
itself. JDOM uses concrete classes rather than interfaces. This
means you can create instances of most of the node types
just by passing an argument or two to a constructor. JDOM
Document objects can also be created by parsing an existing file. This is done by the SAXBuilder class, which relies on
a SAX2 parser such as Xerces. JDOM Document objects can

also be built from existing DOM Document objects through the DOMBuilder class. Moving in another direction, the XMLOutputter class can serialize a JDOM Document object onto a stream. The SAXOutputter class can feed a JDOM Document into a SAX ContentHandler, and the DOMOutputter class can convert a JDOM Document into a DOM Document. After generating the XML file, the software specifications are ready for use. During runtime, if some application level error takes place, it is detected by comparing the usage patterns against the specifications. The process of fault detection is done by the algorithms proposed in the following section.

Abnormalities Detection

The interoperability among systems is commonly achieved through the interchange of XML documents that can represent a great variety of information resources: semi-structured data, database schemas, concept taxonomies, ontologies, etc. Most XML document collections are highly heterogeneous from several viewpoints.

Tag Heterogeneity

The first level of heterogeneity is tag heterogeneity. In our case, it is functional heterogeneity, which means there can be many functions with the same naming conventions and hence it would be difficult to know what is what. Two elements representing the same information can be labeled by tags that are stems (e.g., set_system_parameter and set_system_parameters), that are one substring of the other (e.g., process_authors and process_co-authors), or that are similar according to a given thesaurus (e.g., process_author and process_writer). The second level of heterogeneity is structural heterogeneity, which results in functions with different hierarchical structures. Structural heterogeneity can be produced by the different schemas (i.e., DTDs or XML schemas) behind the XML documents representing functional specifications. Moreover, as schemas can also include optional, alternative, and complex components, structural heterogeneity can appear even for a single program.

We stress the tag and structural heterogeneity of XML document collections as a source to solve the problem detection in software systems. This can lead to search a very large amount of highly heterogeneous documents. In this context, we propose an approach for identifying the portions of documents that are similar to a given pattern. In our context, a pattern is a labeled tree whose labels are those that should be retrieved, preferably with the relationship imposed by the hierarchical structure, in the collection of documents (which, for simplicity, is named the target collection).

Structural Heterogeneity

We develop a two-phase approach where, in the first phase, tags occurring in the pattern are employed for identifying the portions of the target in which the nodes of the pattern appear. Exploiting the ancestor/descendant relationships existing among nodes in the target, sub-trees (named fragments) are extracted having common/similar tags to those in the pattern, but eventually presenting different structures. Moreover, a fifth value is added as a last descendant, which tells which node is the last descendant of that particular node. This increases the performance drastically. The structural similarity between the pattern and the fragments is evaluated as a second step for two purposes: (1) for merging fragments in a region when the region exhibits a higher structural similarity with the pattern than the fragments from which it is originated; and (2) for ranking the identified fragments/regions and producing the result. In this second phase, different similarity measures can be considered, thus accounting for different degrees of document heterogeneity depending on the application domain and on the heterogeneity degree of the target collection. Moreover, a pattern index is created using an array list and sorting it based on level and preorder value that boosts the performance.

The proposed approach is thus highly flexible. However, the problem is how to perform the first step efficiently, that is, how to efficiently identify fragments or portions of the target containing labels similar to those of the pattern without relying on strict structural constraints.

Our approach employs ad-hoc data structures: a similarity-based inverted index (named SII) of the target and a pattern index extracted from SII based on the pattern labels. Through SII, nodes in the target with labels similar to those of the pattern are identified and organized in the levels in which they appear in the target. Fragments are generated by considering the ancestor–descendant relationship among such vertices. Then, identified fragments are combined in regions, allowing for the occurrence of nodes with labels not appearing in the pattern, as described previously. Finally, some heuristics are employed to avoid considering all the possible ways of merging fragments into regions and for the efficient computation of similarity, thus making our approach more efficient without losing precision.

The use of different structural similarity functions taking different structural constraints (e.g., ancestor–descendant and sibling order) into account is discussed. The practical applicability of the approach is finally demonstrated, both in terms of quality of the obtained results and in terms of space and time efficiency, through a comprehensive experimental evaluation. The following conventions are used in order to explain the scenario for better understanding of our approach.

Assumptions

Here are some of the assumptions that we make in order to prove the approach. A tree $T = (V,E)$, where V is a finite set of vertices, E is a binary relation on V that satisfies the following conditions: (1) the root [denoted root(T)] has no parent; (2) every node of the tree except the root has exactly one parent; (3) all nodes are reachable via edges from the root, that is [root(T),v] is the member of E^* for all nodes in $V(E^*$ is the Kleene closure of E). If (u, v) is the member of E, we say that (u, v) is an edge and that u is the parent of v [denoted $P(v)$]. A labeled tree is a tree with which a node labeling function is associated. Given a tree, Table 3.1 reports functions and symbols that we use in this book.

Node order is determined by a preorder traversal, a tree node v is visited and assigned its preorder rank *pre* (v) before its children are recursively traversed from left to right. A post-order traversal is the dual of the preorder traversal: a node v is visited and assigned its post-order rank

Table 3.1   Notations

Symbol	Meaning
Root (T)	Root of T
$v(T)$	Set of vertices of T (i.e., V)
$|V|, |T|$	Cardinality of $v(T)$
Label (v), label (V)	Label associated with a node v and the nodes in V
$\wp(v)$	Parent of vertex v
Pre (v)	Preorder traversal rank of v
Post (v)	Post-order traversal rank of v
Level (v)	Level in which v appears in T
Pos (v)	Left-to-right position at which v appears among its siblings
desc (v)	Set of descendant of v ($desc(v) = \{u|(v, u)) \in E\}$)
nca (v, u)	Nearest common ancestor of v and u
d (v)	Distance of node v from the root
d^{max}	Maximal distance from the root ($d^{max} = \max_{\in v(v)d(v)}$)

post (*v*) after all its children have been traversed from right to left. The *level* of a node *v* in the tree is defined, as usual, by stating that the level of the root is 1, and that the level of any other node is the successor of the level of its parent. The position of a node *v* among its siblings is defined as the left-to-right position at which *v* appears among the nodes whose father is the father of *v*. Each *v* is coupled with a *quadruple* (*pre*(*v*), *post*(*v*), *level*(*v*), *pos*(*v*)) as shown in Figure 3.4 *pre* (*v*) is used as node identifier.

Pre- and post-ranking can also be used to efficiently characterize the descendants *u* of *v*. A node *u* is a descendant of *v*, *v* ∈ *desc*(*u*), iff *pre* (*v*) < *pre* (*u*) ∧ *post* (*u*) < *post* (*v*). Given a tree *T* = (*V*,*E*) and two nodes *u*, *v* ∈ *V*, the nearest common ancestor of *u* and *v*, *nca* (*u*,*v*), is the common ancestor of *u* and *v* such that any other common ancestor of *u* and *v* is an ancestor of *w*. Note that $d^{max} = max_\varepsilon \vee (v)d(v)$ if u is a descendant of v. The distance *d*(*v*) of a node *v* from the root, which coincides with its preorder rank, amounts to the number of nodes traversed in moving from the root to the node in the preorder traversal. The maximal distance d^{max} corresponds to the number of nodes in the tree, that is, |T|. d^{max} also corresponds to the post-order rank of the root.

Similarity Matrices

Label Similarity

Labels can be similar or dissimilar depending on the adopted criteria of comparison specified by means of functions. In this chapter, the following similarity functions have been considered, even if other ones can be easily integrated:

■ Case Insensitive Similarity Functions S_{ci}^t: Two tags are similar if their differences depend only on the case; for example, "programmer" and "Programmer" are similar.

- Stemming Functions S_{st}^t: Two tags are similar if one is the stem of the other; for example, "programmer" and "programmers" are similar.
- Edit Distance Function S_{ed}^t: Two tags are similar if their edit distance is less than a pre-fixed threshold; for example, "programmer" and "prog" are similar.
- Substring Similarity Function S_{ss}^t: Two tags are similar if one is contained in the second one; for example, "coder" and "class-coder" are similar.
- Semantic Similarity Function S_Q^t: Two tags are similar if they are synonyms of each other; for example, programmer and coder are similar.
- Tree Edit Distance Similarity Function S_{ed}^t: Two tags are similar if their tree edit distance is similar to each other, e.g., number of hops one needs to move from one node to anotherr is called the tree edit distance between these two nodes.

A pattern is a labeled tree. The pattern is a tree representation of the user interest and can correspond to a collection of navigational expressions on the target tree. Consider the pattern in Figure 3.5. Intuitively, the pattern expresses an interest in portions of a software program related to functions, their names, and the classes to which they are belong. Possible matches for

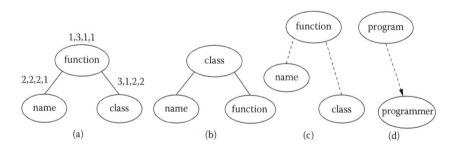

(a)   (b)   (c)   (d)

Figure 3.5   A pre/post-order rank. (a) A matching fragment with a different order, (b) missing levels, (c) missing elements, (d) semantic similarity.

this pattern are reported in Figure 3.5(a–d). The matching tree in Figure 3.5(b) contains similar labels but at different positions, whereas the one in Figure 3.5(c) contains similar labels but at different levels. Finally, the matching tree in Figure 3.5(d) misses an element and the two elements appear at different levels.

The target is a set of heterogeneous pattern strings/sub-trees in a source. The target is conveniently represented as a tree with a dummy root labeled db and whose sub-elements are the documents of the source. This representation relies on the common model adopted by native XML databases and simplifies the adopted notations. An example of target is shown in Figure 3.6. The dummy root has preorder rank 0 and is at level 0 in the tree.

The target can be defined mathematically as follows.

Let $\{T_1, \ldots T_n\}$ be a collection of trees, where $T_i = (V_i, E_i)$, $1 \le i \le n$. A target is a tree $T = (V, E)$ such that:

$$V = \bigcup_{i=1}^{n} X_i \grave{E}\{r\} \text{ and } r \notin \bigcup_{i=1}^{n} V_i,$$

$$E = \bigcup_{i=1}^{n} E_i \cup \{(r, root(T_i)), 1 \le i \le n\}$$

$$label(r) = db$$

The basic building blocks of our approach are fragments. Given a pattern P and a target T, a fragment is a sub-tree of T, belonging to a single document of the target, in which only nodes with labels similar to those in P are considered. Two vertices u, v belong to a fragment for a pattern iff their labels as well as the label of their nearest common ancestor similarly belong to the labels in the pattern. Edges in the fragment correspond either to a direct edge in the target (father–children relationship) or to a path in the target (ancestor–descendant

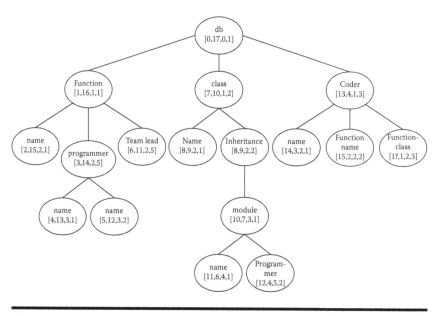

Figure 3.6   Schema under consideration.

relationship). Several edges in the target, indeed, can be col-
lapsed in a single edge in the fragment by "skipping" nodes
that are not included in the fragment because their labels do
not similarly belong to those in the pattern.

A fragment F of a target $T = (V_T, E_T)$ for a pattern P is a sub-
tree (V_T, E_T) of T for which the following properties hold:

V_F is the maximal subset of V_T such that root (T) is not the
   member of V_F and $\forall u, v \in V_F$, *label, label* (v), and *label*
   $(nca(u,v))$ α label $(V(P))$;
for each $v \in V_F$, $nca(root(F), v) = root(F)$;
$E_F = \{(u,v) | u,v \in V_F \wedge (u,v) \in E_T \wedge (\nexists w \in V_F, w \neq u,v, s.t.(u,w)$
   $\in E_T \wedge (w,v) \in E_T)\}$

Consider the pattern in Figure 3.5(a) and the target in
Figure 3.6. By considering all the label similarity functions, the
corresponding four fragments are shown in Figure 3.7. Labels in
the first fragment are exactly the same appearing in the pattern.
By contrast, the others require exploitation of the substring and

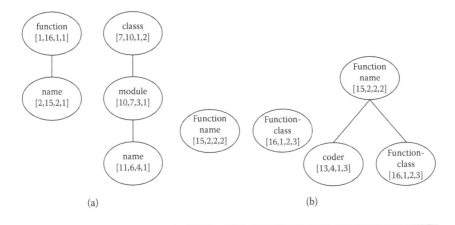

Figure 3.7   Fragments and generated regions.

semantic similarity-based functions. For instance, the second tree contains function as label, and function is semantically similar to program. Similarly, the third and fourth trees contain node labels that are similar, exploiting the substring function to labels name and class in the pattern, respectively. The second tree provides an example of fragment in which a node of the original tree [i.e., node (9, 8, 2, 2) labeled by inheritance] is not included.

Starting from fragments, regions are introduced as a combination of fragments rooted at the nearest common ancestor in the target. Two fragments can be merged in a region only if they belong to the same document. In other words, the common root of the two fragments is not the db node of the target.

Consider the tree *T* rooted at node *n* (13,4,1,3) in Figure 3.7. It has two sub-trees (the ones containing elements function-name and function-class) that are fragments with respect to the pattern in Figure 3.5. Although *n* is not (part of) a fragment, the sub-tree consisting of *n* and its fragment sub-trees could have a higher similarity with the pattern tree in Figure 3.5 than its sub-trees separately. Therefore, combining fragments into regions may lead to sub-trees with higher similarities.

A region can be a single fragment or it can be obtained by merging different fragments in a single sub-tree whose root is the nearest common ancestor of the fragments. Thus, while all fragment labels similarly belong to those in the pattern, a region can contain labels not similarly belonging to pattern labels.

Regions can be defined as follows:

Let $F_P(T)$ be the set of fragments identified between a pattern P and a target T. The corresponding set of regions $R_P(T)$ is inductively defined as follows:

$$F_P(T) \subseteq R_{P(T)}$$

For each $F = (V_T, E_T) \in F_P T$ and for each
$$R = (V_R, E_R) \in R_P(T)s.t.label\big(nca\,(root(F), root(R))\big) \neq$$
$$db, S = (E_S, V_S) \in R_P(T); where : -root(S) =$$
$$nca\,(root(F), root(R)), -V_S = V_F \cup V_R \cup \{root(S)\}, -E_S =$$
$$E_F \cup E_R \cup \{(root(S), root(F)), (root(S), root(R))\}$$

The notions of level of a node and distance between two nodes, when applied to regions, refer to the corresponding notions in the original target tree. More specifically, they refer to the target sub-tree whose root is the region root and whose leaves are the region leaves. We will refer to this tree as the target sub-tree covered by the region. This sub-tree may contain additional internal nodes that are not included in the region because their labels do not appear in the pattern. Specifically, internal nodes are included in the covered sub-tree either if they are in the path from the region root to some region node or if they are internal siblings of two nodes in the covered tree (i.e., right-sibling of one of them and left-sibling of the other). We can define covered sub-trees as follows. Let T be a target and R be a region on it. The sub-tree of T covered by R, denoted as $C(R)$, is the sub-tree of T such as:

■ $V_C(R) \subseteq V_T$ is inductively defined as follows:
  - $V_R \subseteq V_C(R)$
  - $\forall_v \in V_T$ such that $\exists_{u,} w \in V_{c(R)}$ and $(u, v), (v, w)$
    $\in V_T, v \in V_{c(R)}$

– $\forall_v \in V_T$ such that $\exists u,w \in V_{c(R)}$ and (f, u), (f, v), (f, w) $\in V_T$, u is the left sibling of v, and w is the right sibling of v, $v \in V_T$

■ $V_c(R) = u,v \,|\, u,v \in E_T s.t. u,v \in V_c(R)$

Figure 3.8 presents different parts of a target where black nodes identify region nodes and gray nodes together with black nodes form the covered sub-tree.

Pattern Similarity

In this section, we present the foundation of our two-phase approach to identify target regions similar to a pattern. We first identify the possible matches between the vertices in the pattern and those in the region having similar labels, without exploiting the hierarchical structure of the tree. Then, the hierarchical structure is taken into account to select, among the possible matches, those that are structurally more similar. Specifically, after having introduced the definition of mapping, we propose three different similarity measures that combine structure and tag matching.

A mapping between a pattern and a region is a relationship among their elements that takes the tags used in the programs into account.

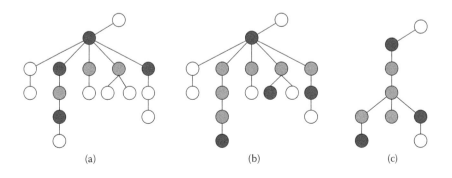

(a)   (b)   (c)

Figure 3.8   Covered sub-trees in a target.

Let *P* be a pattern and *R* be a region. A mapping *M* is a partial injective function between the vertices of *P* and those of *R* such that $\forall x_p \in V(P), M(x_p) \neq \perp \Rightarrow label(x_p) \cong label(M(x_p))$. Figure 3.9 reports the pattern *P* of Figure 3.6 in the center and three target regions, R1,R2,R3. Dashed lines represent a mapping among the vertices of the pattern and those of each region. Several mappings can be established between a pattern and a region. The best one will be selected by means of a similarity measure that evaluates the degree of similarity between the two structures relying on the degree of similarity of their matching vertices. Let *M* be a mapping between a pattern *P* and a region *R*, and let *Sim* be a vertex similarity function. The evaluation of *M* is:

$$Eval(M) = \frac{\sum_{i=1}^{n} X_{p \in V(P) s.t. M(x_p) \neq \perp} Sim(x_p, M(x_p))}{|V(P)|}$$

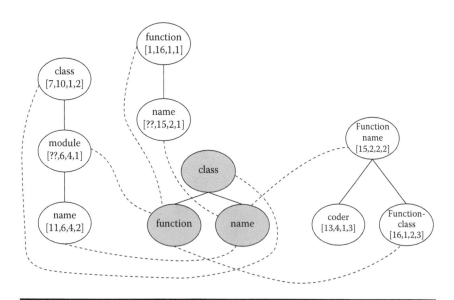

Figure 3.9   Mapping between patterns and different regions.

Similarity between a pattern and a region is then defined as the maximal evaluation among the mappings that can be determined between the pattern and the region.

Let M be the set of mappings between a pattern P and a region R. The similarity between R and P is defined as

$$Sim(P, R) = \max_{m \in M} Eval(M)$$

or

$$Sim(P, R) = \max_{m \in M} .Eval(M)$$

$$= \frac{\sum_{i=1}^{n} X_{p \in V(P) s.t. M(x_p) \neq \perp} Sim(x_p, M(x_p))}{|V(P)|}$$

Vertices Similarity

The similarity between matching vertices is a tricky task as we present an approach for computing the similarity between a pair of matching vertices that takes the structure into account.

In the first approach, similarity only depends on node labels. Similarity is 1 if labels are identical, whereas a pre-fixed penalty d is applied if labels are similar. If they are not similar, similarity is 0. The match-based similarity can be defined as:

$$Sim_M(x_p, X_r) = \begin{cases} 1 & if & label(x_p) = label(x_r) \\ 1 - \delta & if & label(x_p) \cong label(x_r) \\ 0 & & otherwise \end{cases}$$

In the second approach, the match-based similarity is combined with the evaluation of the level at which x_p and $M(x_p)$ appear in the pattern and in the region. Whenever they appear in the same level, their similarity is equal to the similarity computed by the first approach. Otherwise, their similarity linearly decreases as the number of levels of differ-ence increases. We recall that levels in the region refer to the levels in the target sub-tree covered by the region. It can be defined as:

$$Sim_L(x_p, x_r) = Sim_M(x_p, x_r) - \frac{level_p(x_p) - level_R(x_r)}{\max(level(P), level(R))}$$

The similar is 0 if the obtained value is below 0.

The distance based is computed by taking the distance of nodes x_p and $M(x_p)$ with respect to their roots into account. Thus, in this case, the similarity is the highest only when the two nodes are in the same position in the pattern and in the region. We recall that distances in the region refer to the distances in the target sub tree covered by the region com-puted through the recursive function d_R in Figure 3.10. Given a region R, $v_1, \ldots v_n$ are the vertices of R ordered according to their pre-order rank in the target T. In the figure, we report the computation of the distance for the root of $R(V_1)$ and for generic vertex of E that is not the root.

Considering the approach mentioned above, we calculate the best matched fragments from the usage log and match it against the specification log. The mismatches are forwarded to the policy engine that finds the related exceptions and solu-tions to propose a healing policy.

$$d_R(v_i) = 1$$

$$d_R(v_i) = d_R(v_{i-1}) + \begin{cases} level(v_i) - level(v_{i-1})v_i \in desc(v_{1-1}) \\ pos(v_i) - pos(v_{i-1})v_i sibling\ of\ (v_{i-1}) \\ pos(av_i) - pos\ pos(av_{i-1}) + level(v_i) \\ \quad - level\ pos(av_i) otherwise \end{cases}$$

Figure 3.10   Distance of vertex in a Region R.

Tree Edit Distance Similarity

The tree-edit distance is a natural metric for correcting and discovering approximate matches in XML document collections. We use a deterministic approach for smaller size trees in order to keep the performance within acceptable range.

We use the following procedure to calculate the tree edit distance based similarity.

$$\|V(S) - V(T)\|_1 = \sum |V(S)[i] - V(T)[i]|$$

The implementation of the solutions is shown in Chapter 5.

Chapter 4

Policy Engine

Introduction

The increasing cost of maintaining IT systems relative to the cost of purchasing those systems has led to an increased emphasis on use of autonomic computing principles to reduce the need for skilled labor for its use, support, maintenance, and management. This entails that autonomic computing systems be able to dynamically adapt themselves relatively easily with respect to changing business conditions and objectives. That is why the importance of policy-based system has gradually been recognized in autonomic computing systems. Moreover, policy-based technologies promise to reduce the burden of managing large-scale computing systems by freeing human administrators from direct manipulation of underlying devices and systems, and by providing systematic means to create, modify, and distribute policies. That is the reason we develop a policy engine on top of our policy engine to improve the functionality.

The prevalence of networked systems and agile applications in business environments has created an immense challenge for IT administrators. As networks become ever more hetero-geneous and connected to each other, scalable and distributed management of IT infrastructure becomes imperative. In order to address this issue, policy-based management has been pro-posed in recent years.[1-3] Instead of running customized scripts and manually configuring and auditing networked devices and applications, policy-based management allows IT admin-istrator to specify high-level directives or policies for various management tasks such as network planning, problem detec-tion, security, and quality of service (QoS) provisions. Recently, policy-based frameworks have been successfully implemented in various domains, such as storage area networks,[4] networked systems,[5] database management,[6] security policy like authoriza-tion and access control,[7,8] and QoS provisions.

Policies can be specified in many different ways and mul-tiple approaches have been suggested for different applica-tion domains.[9] KAos[10] is a collection of componentized policy and domain management services originally designed for governing software agent behavior and then adapted to grid computing. Rei is a policy framework that integrates support for policy specification, analysis, and reasoning in pervasive computing applications.[11] Ponder is a declarative, object-oriented language that supports the specification of several types of management policies for distributed object systems.[12] Many proposals have also focused on specific system man-agement tasks. In Reference 13, the authors have presented a set of algorithms for detecting anomalies in firewall poli-cies. In Reference 14, the authors have presented a policy-based resource management tool for capacity planning using MPLS in the IP networks to enable differentiated services. In Reference 7, the authors have proposed a policy-based system to manage the QoS of video stream application. However, as the heterogeneity of devices increases and the connectivity of networks grows, an integrated approach to policy man-agement is needed. In addition, many researchers proposed a variety of adaptation methods for software components in the area of software development methodology, emphasizing on extensibility and adaptability.[15,16] However, the application of those solutions in a real-time application decreases performance, which is the motivation of our work. In order to answer this weak point, the techniques of policy-based component development are proposed.[15] For extensibility and adaptability of components, the techniques separate business variability[15] from the component's internal code by keeping separate rules. Upon the occurrence of requirement changes, a new requirement can be satisfied with changes in rules without changes in components. Currently, security policy-based approaches have been proposed for authorization and access control in dynamic and collaborating Web service environments.[7,8] Many of these researchers adopt the eXtensible Access Control Markup Language (XACML)[17] as a policy description language.[18–20] It provides a syntax (in XML) to define action (request) rules for subjects (users) and targets (resources). XACML describes both an access control policy language and request/response messages. The policy language is used to express access control policies (who can do what, where, and when). The request/response language expresses queries about whether a particular access should be allowed (requests) and describes answers to those queries (response).

However, all of these policy-related technologies usually need some additional script languages to describe policy expression, which has the limitation in expressing complex business policies and gives additional burden to policy writers (i.e., they have to learn a difficult policy language). Moreover, since these script-based policies usually spend a lot of time on parsing the script language during run time, they are not suitable to system environments that require high performance and low computing power such as real-time systems and the network management in a sensor network. Therefore, we need policy-related technology that provides rich policy expression without additional burden to policy writers in design time and high performance in run time.

We propose a model-integrated compilation-based policy system for performance enhancement and improving policy expression to cope with a dynamic system requiring runtime adjustments. Unlike interpretation-based policy engines proposed as contemporary solutions, our policy engine does not require any additional script language for expressing rules resulting in better performance in terms of time compilation and overall performance. We use Java language in order to create/modify rules instead of scripting languages. It gives us the facility of standardized syntax as well. It separates the condition from action during run time, which makes rule notification easier and quicker. However, since a policy writer usually does not know Java language, it might be difficult in describing policies. In order to solve this problem, we choose to adopt a model-integrated computing environment, which enables the policy writers to describe their policies in a modeling environment and then automatically generate the condition and action Java codes that are used by the compilation-based policy engine. Moreover, the solution we propose is able to use the current existing libraries for condition/action codes of rules in legacy systems, such as string, number, and logical expression so that it may not only express complex condition or action statements but also easily integrate the existing systems developed in Java language. In agile business computing environments, computing systems have become highly capricious and complex.

Software Architecture

Dynamically Adjustable Policy Engine

In order to apply a changing rule to a dynamically adjustable policy engine, it is an integral proposition that the policy engine should be adaptable to tackle with regular updates and changes. The main procedure of our dynamically adjustable policy engine is that a policy writer describes a condition and obligation part of a policy expression in a modeling environment and then automatically generates condition and obligation Java codes by using an interpreter of the modeling environment. The condition code and action code of a rule expression converts into a condition and action object with a hook method and puts them into an object pool. After finding a specific rule, our policy engine takes the condition and action objects specified by the policy's configuration from the object pool for policy execution. Processing a sample scenario is introduced in the following subsections.

A Sample Scenario of the Dynamically Adjustable Rules

Figure 4.1 shows the application example of customer's credit rule. Suppose that there is a rule of the customer's credit in an import and export business domain.

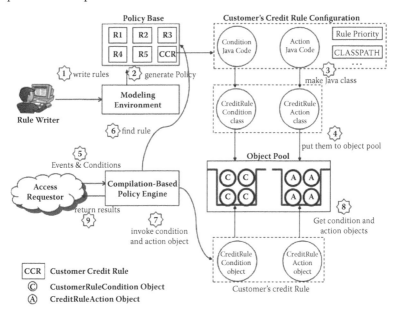

Figure 4.1   Application example of customer's credit rule.

Let us consider a simple credit rule: "If a customer's credit limit is greater than the invoice amount and the status of the invoice is 'unpaid,' then the credit limit decreases by taking off the invoice amount and the status of the invoice becomes 'paid.'"

In this scenario, the process of applying the dynamically adjustable rules can be divided into three phases: (1) the rule expression phase, (2) the rule initialization phase, and (3) the rule execution phase. Figure 4.2 shows the architecture of a model-integrated compilation-based policy system. A policy writer needs to describe the meta-model and user model of a policy application and then automatically generate the condition and obligation Java codes by using an interpreter. Those are built using the generic modeling environment (GME),

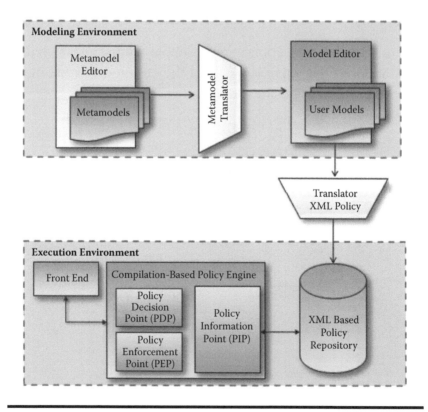

Figure 4.2 Model-integrated compilation-based policy system architecture.

which is part of the model integrated computing tool suite.[17,18] GME is a configurable toolkit for creating domain-specific modeling and program synthesis environments. The configuration is accomplished through meta-models specifying the modeling paradigm (modeling language) of the application domain. The modeling paradigm contains all the syntactic, semantic, and presentation information regarding the domain—which concepts will be used to construct models, what relationships may exist among those concepts, how the concepts may be organized and viewed by the modeler, and rules governing the construction of models. The meta-models specifying the modeling paradigm are used to automatically generate the target domain-specific environment. The generated domain-specific environment is then used to build user domain models that are stored in a model database. These domain models are used to automatically generate the application or to synthesize input to different COTS analysis tools. This process is called model interpretation. In GME, the interpreter does the model interpretation.

Figure 4.3 shows a meta-model for a user domain model. The meta-model is made with GME and it usually describes the semantics, structure, and constraints of policy concepts. Many policies can be depicted in a policy designer sheet. A policy can include zero or more rules and include many targets that have zero or more subject, resource, and environment. We omit the details here. In addition, a policy can have zero or more obligations, which are the kind of jobs that need to be done if the result of a condition is true. A rule has many conditions that are evaluated using value matching between incoming parameter values and a target's values. We can make a policy user model by using the previous meta-model.

Figure 4.4 shows the condition part of the customer credit policy user model. In order to evaluate the condition of the customer credit rule, we apply two functions: Check Credit Limit function and Check State of Invoice function. The Credit Limit function compares the values between the customer

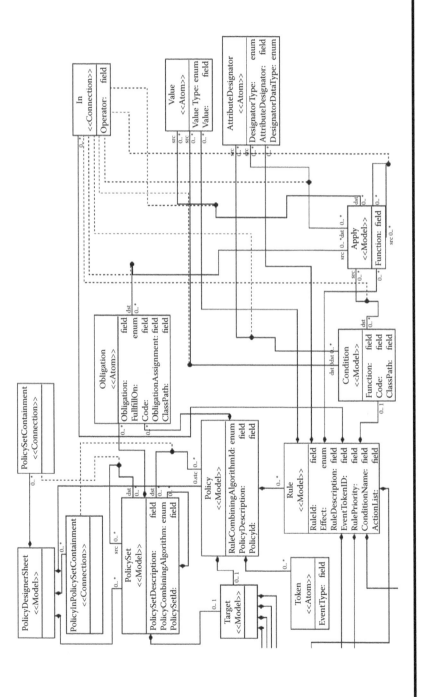

Figure 4.3   A policy meta-model made by GME.

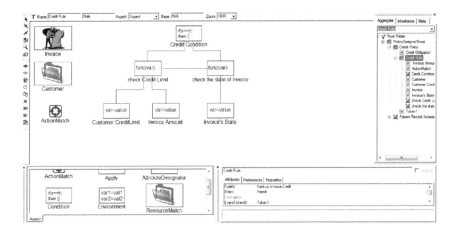

Figure 4.4   A condition part of the customer credit policy user model.

credit and the requested amount of invoice. The Check State of Invoice function checks whether the state of the invoice is unpaid or paid. We use the operator attribute of "in connection" in order to specify the relationship between two functions. For example, if the value of operation attribute is specified as an "And," both of two functions are needed to be true in order to satisfy the Credit Condition of Figure 4.4.

After writing the policy user model, we need to implement an interpreter in order to interpret the policy user model. A coder of the interpreter accesses the entities and attributes of the customer credit policy user model, and then generates an XML-based policy file like that in Figure 4.5. There are condition and obligation Java codes in the code element of the XML-based policy file. These Java codes are generated by the interpreter codes from the customer credit policy user model. The interpreter code generates many configuration elements required to execute the customer credit policy.

During the policy initialization phase, the policy engine gets action and obligation codes from the code element of Figure 4.5, compiles them, makes instance of the classes, and deploys them to the object pool. During the rule execution phase, if an access requestor sends request event messages to

```
<Configuration>
    <JavaClassPATH>E:\2006\development_workspace\RuleEngine\bin </JavaClassPATH>
    <JavaCompilerPATH>E:\2006\development_workspace\RuleEngine\bin </JavaCompilerPATH>

    <RuleDefinition>
        <ruleIdentitifier>Konkuk:invoice:Credit </ruleIdentitifier>
        <ruleName>Credit Rule </ruleName>
        <rulePriority>5</rulePriority>
        <eventTokenID>Token1</eventTokenID>
        <conditionName>Credit Condition</conditionName>
        <ActionList>
            <ActionName>Credit Action</ActionName>
        </ActionList>
    </RuleDefinition>

    <ConditionDefinition>
        <ConditionName>Credit Condition</ConditionName>
        <CLASSPATH>.;E:\2006\development_workspace\RuleEngine\bin </CLASSPATH>
        <Code>
            Customer cs =(Customer)(events.getDataField("Customer"));
            Invoice iv =(Invoice)(events.getDataField("Invoice"));
            if((cs.getCreditLimit() > iv.getAmount() && iv.getStatus().equals("unpaid")) {
                return true;
            }
            else {
                return false;
            }
        </Code>
    </ConditionDefinition>

    <ActionDefinition>
        <ActionName>Credit Condition</ActionName>
        <CLASSPATH>.;E:\2006\development_workspace\RuleEngine\bin </CLASSPATH>
        <Code>
            Customer cs =(Customer)(events.getDataField("Customer"));
            Invoice iv =(Invoice)(events.getDataField("Invoice"));
            int decreasedAmount = (cs.getCreditLimit() - iv.getAmount());
            cs.setCreditLimit(decreasedAmount);
            iv.setStatus("paid");
        </Code>
    </ActionDefinition>

    <TokenDefinition>
        <TokenID>Token1</TokenID>
        <EventTyp>KONKUK:INVOIDE:CREDIT</EventTyp>
    </TokenDefinition>
</Configuration>
```

Figure 4.5   XML-based policy file for customer's credit rule expression.

the policy engine, the policy engine extracts the event identifier from the request event message. The policy engine finds
the rule from a rule base by matching the event identifier.
The policy engine takes condition and action objects from the
object pool and invokes the hook method of the condition
and action objects. The rule identifier is the unique name for
finding the specified rule and the rule priority specifies the
order of executing rules. It is also possible to use the existing
libraries specified in CLASSPATH. If necessary, a rule writer
can write multiple action codes for a rule.

Code Generation and Operation in the Policy Engine

In order to generate condition and action classes, the policy
engine uses a template method pattern. Figure 4.6 shows the
class diagram for applying the temple method pattern to our

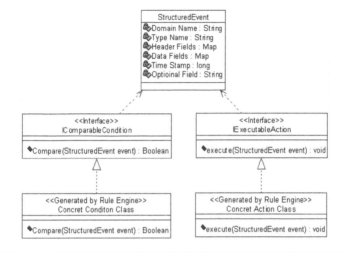

Figure 4.6   Condition and action class generation using template method pattern.

policy engine. The names of the hook method for condition and action classes are "Compare" and "Execute," respectively. Figure 4.7 shows condition or action codes generated automatically through the template method pattern. The condition and

```
import dmics.ens.rule.*;
import dmics.ens.notification.*;

public class CreditRuleCondition implements IComparableCondition {
    public boolean compare(StructuredEvent events) {

        Customer cs =(Customer)(events.getDataField("Customer"));
        Invoice iv =(Invoice)(events.getDataField("Invoice"));
        if((cs.getCreditLimit() > iv.getAmount()) && iv.getStatus().equals("unpaid")) {
            return true;
        }
        else {
            return false;
        }
    }
}
```
An Example of the Condition Code

```
import dmics.ens.rule.*;
import dmics.ens.notification.*;

public class CreditRuleAction implements IExecutableAction {
    public void execute(StructuredEvent events)   {
        Customer cs =(Customer)(events.getDataField("Customer"));
        Invoice iv =(Invoice)(events.getDataField("Invoice"));
        int decreasedAmount = (cs.getCreditLimit() - iv.getAmount());
        cs.setCreditLimit(decreasedAmount);
        iv.setStatus("paid");
    }
}
```
An Example of the Action Code

Figure 4.7   Condition and action code generation for customer's credit rule.

action objects are made from the CreditRuleCondition and the CreditRuleAction class and put into an object pool to be used for executing the rule. When a rule application sends request events for rule execution to the policy engine, the policy engine extracts the event identifier from the request event message. The event identifier is the string of "domain name: task identifier: rule name". The policy engine finds the rule from a rule base by matching the event identifier. The rule matched has rule configuration, such as rule identifier, rule name, condition or action class name, and rule priority. The policy engine takes condition and action objects from the object pool and invokes the hook method of the condition and action objects.

Software Architecture of the Policy Engine

In the previous section, we studied a sample scenario with processing flow. This section introduces the architecture of the dynamically adjustable policy engine, which is operated based on compilation. In addition, we present flow of the initialization process in the policy engine and execution process of rules. In Figure 4.8, we show the software architecture of the proposed policy engine. The policy engine is mainly comprised of three parts: Admin Console, Rule Repository, and Core Modules. The Admin Console is the toolkit for expressing and managing rules. The Rule Repository saves the XML-based rule information expressed by the toolkit. The Core Modules are in charge of finding, paring, and executing rules. There are a number of modules in the Core Modules. The responsibility of the policy engine is to receive request messages from a client and to execute rules. To find an appropriate rule, it sends the request message to the Rule Parser. The Rule Parser extracts the event identifier from the request message, compares it with the event identifier of a parsing table, and finds the rule. The event identifier is the string of "domain name: task identifier: rule name". After finding the rule, the policy engine knows the names of condition and action objects from

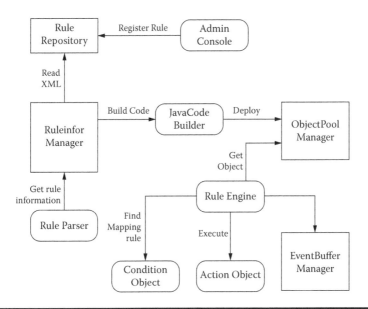

Figure 4.8   Software architecture of the proposed policy engine.

the configuration of a rule and has the references of them from the ObjectPool Manager.

The Rule Parser is responsible for finding rules. The ObjectPool Manager manages the condition and action objects specified in rule expression. The RuleInfor Manager performs CRUD (Create, Read, Update, and Delete) action on the Rule Repository. The JavaCode Builder makes Java source files, compiles them, makes instances of the classes, and deploys them to the object pool. The Condition and Action Objects are the objects made from condition and action codes of rule expression.

The policy engine is required to initialize before executing rules. In Figure 4.9, we show the collaboration diagram to show the flow of the process for policy engine initialization. The policy engine sends an initialization request to the RuleInfor Manager. The RuleInfor Manager reads rule information from the Rule Repository and saves it to a buffer. Recursively, the RuleInfor Manager extracts condition and action codes of rules, makes object instances, and deploys to

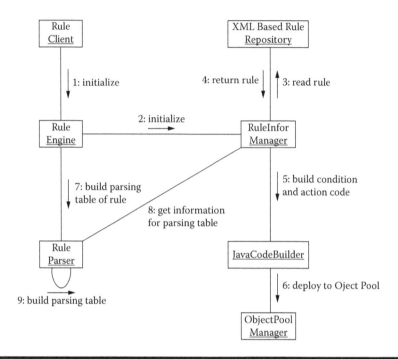

Figure 4.9   Process flow for rule initialization.

the object pool through the ObjectPool Manager. After the policy engine initializes the condition and action parts of a rule, it calls the Rule Parser for building a parsing table. The Rule Parser gets a pair of rule identifiers and names from the RuleInfor Manager, and builds the parsing table with them for finding appropriate rules.

Figure 4.10 presents the collaboration diagram to show the flow for rule execution. A client sends request messages to the policy engine. The policy engine saves it to a buffer through the EventBuffer Manager and then gets the request message with highest priority from the EventBuffer Manager.

The policy engine calls the Rule Parser for finding the rule matched with the rule identifier. The Rule Parser searches the parsing table to find appropriate rules. After finding the rule, the policy engine calls the ObjectPool Manager to get the condition and action objects specified in the founded rule and then calls the "Compare" hook method of the condition object.

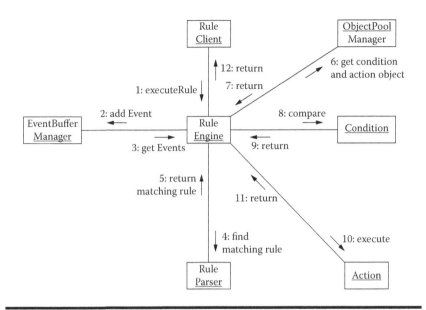

Figure 4.10   Process flow for rule execution.

If the result of invocation of the condition object is true, the policy engine calls the "Execute" hook method of the action object. If a rule has many action objects, the policy engine calls them according to the order of the action object specified in rule expression. The policy engine also supports the forward-chaining rule execution. It allows the action of one rule to cause the condition of other rules.

Performance of the Policy Engine

In this section, we show the experimental performance results of the compilation-based policy engine proposed in this chapter. We used the Microsoft 2003 server for operation system, WebLogic 6.1 with SP 7 for web application server, and Oracle 9i for relational database. As for load generation, WebBench 5.0 tool was employed. Transactions per second (TPS) and execution time were used for the metric of performance measurement. For performance comparison in J2EE environment, we used a servlet object as a client of the policy engine.

Experimental Environment

Before showing the performance results, we introduce the workloads that were used in the experiments. Generally, business rules are classified into business process rules and business domain rules. Business domain rules define the characteristics of variability and the variability methods that analyze these characteristics for an object. Business process rules define the occupation type, sequence, and processing condition, which is necessary to process an operation. In the business process rule, the variability regulations for process flows are defined as the business process rules. Table 4.1 shows the workload configuration for experiments. Among the five rules, two rules are the business process rules and the other two rules are the business domain rules. In an e-business environment, as the business domain rules are more frequently used than the business process rules, we give more weight to the business domain rules.

The "Customer Age" rule measures a customer's age according to problem request. The "Interest Calculation" rule calculates interest according to the interest rates. The "Role Checking" rule specifies the assertion of the "An authorized user can access certain resources." The policy engine takes role information from the profiles of the customer and decides whether the requesting jobs are accepted or not.

Table 4.1 Workload for Experiments

Index	Rule Name	Rule Type	Weight (%)
1	Log-In	—	5
2	Customer Credit	Process Rule	15
3	Customer Age	Domain Rule	30
4	Interest Calculation	Process Rule	15
5	Role Checking	Domain Rule	35

Performance Comparison

The performance of the proposed policy engine is compared with Java Policy Engine API (JSR-94) in Figure 4.11. The proposed policy engine achieved 395 TPS in maximum workload, while JSR-94 achieved 150 TPS in maximum. The proposed policy engine processes 245 more TPS than JSR-94. We believe that the policy engine proposed achieved 2.5 times better performance than JSR-94 because of its special emphasis on features like ease in extensibility and high levels of adjustability for rules that are used in a system. In order to compare performances of sub-modules of the policy engine, Figure 4.12

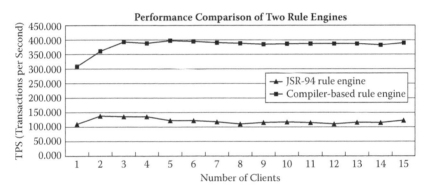

Figure 4.11   Performance comparisons with JSR-94.

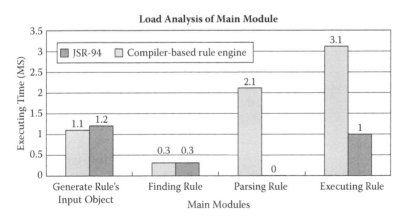

Figure 4.12   Comparison of load of two policy engines.

shows load analysis of two policy engines. Since the proposed policy engine operates on compilation-based rule processing, performance in the module of generating objects may take a long execution time, but there is not a big difference in performance. Moreover, the proposed policy engine achieves better performance results in parsing and executing rules. This is because it divides the condition and action class into separate parts, which gives ease at run time when rules are called in an object pool. Moreover, one does not have to define separate condition statements for multiple actions. The proposed policy engine provides the facility of defining more than one execution for one condition, which can help in fault tolerance in a hybrid environment.

Feature Comparison

In Table 4.2, we compare the features of the two policy engines. In contrast to JSR-94, the proposed policy engine expresses each business rule by a business task unit. If there are one or more rules in a task, each rule is categorized in a unique rule name.

Table 4.2   Feature Comparison between the two policy engines

	JSR-94 Rule Engine	The Proposed Rule Engine
Performance (Max TPS)	150 TPS	395 TPS (2.5 times better performance)
Rule Expression	A rule expression is confined by the JESS script rule language	Need to learn Java language Can express complex business rules using the Java language
Reusability of Existing Libraries	Impossible	Possible by using the CLASSPATH in rule expression

(continued)

Table 4.2   Feature Comparison between the two policy engines (continued)

	JSR-94 Rule Engine	The Proposed Rule Engine
Integration of Existing System Using Rule Engine	Needs additional rule expression for integrating existing systems	Easier to integrate with existing systems in Java language
Easy to Learn	An application domain expert is easier to write rules. Needs to learn additional script-based rule language	Any Java coder can be easier to write rules Learning additional rule language is not required
Dynamic Change of Business Rules	Possible	Possible (an object pool mechanism of condition and action objects can make dynamic change of rules)
Separation of Condition and Action Parts	No	Yes, the condition and action part of rules are separated so that the updates are easier to manage and multiple actions could be taken against one condition
Ease of Embedment	Low	High
The Condition/ Action Dependability	Yes, causes rule evaluation to block until a condition becomes true or an event is raised	No, since conditions and events are "physically" separate from each other, it gives the proposed engine an edge on time constraint

The proposed policy engine uses Java language for writing business rules without using any additional script languages for expressing rules. Although it might seem odd to assume that the user must have knowledge of Java language, we foresee that the business rules, when converted into Java language, eliminate the fuzziness and bring clarity to the conditions and actions. Moreover, syntax of Java is the same everywhere in the world so it would be easier to embed the proposed policy engine into applications facing diverse environment. However, we aim to build a GUI-based front end in the policy engine proposed in this book as future work.

Whenever executing each business rule in the proposed policy engine, the step for matching rule conditions is not required. In other words, after finding the required business rule from a rule base, the proposed policy engine executes it without parsing the rule and matching the rule conditions due to Java-based rule expression. The proposed policy engine converts the condition and action codes of a rule into condition and action objects, respectively, and puts it into an object pool for improving performance and dynamic changeability. Thus, it can execute the newly changing business rule without restarting itself.

As business applications become complex and changeable, a rule-based mechanism is needed for automatic adaptive computing as well as trustworthy and easy maintenance. For this purpose, we propose a compilation-based policy engine that can easily express business rules in Java codes. It does not need additional script language for expressing rules. It can create and execute condition and action objects at run time. Moreover, it can use existing libraries for condition or action codes of rules (i.e., string, number, and logical expression) so that not only can it express complex condition or action statements, but also it can easily integrate the existing systems developed in Java. The compilation-based policy engine proposed in this book shows better performance than JSR-94, a generally used interpretation-based policy engine. According to our experiments, the proposed policy engine processes 245 more TPS than does JSR-94. We intend to test the performance of the policy engine proposed in this research with different weights and in different conditions. Not only will this give us a better idea about the working capacity of the outcome of this research, but also it will give clear application for this policy engine. Moreover, we intend to develop a GUI that could assist the users who have limited knowledge of Java in operating with this policy engine.

Novel Rule Methodology

The rules, often referred to as policies, need frequent changes mainly because of ever-changing situations in the modern world. These frequent changes can lead to expensive abnormalities and complexities. In order to avoid these errors, a mechanism that could ensure consistency, provide easy access to the user, assist the user in guaranteeing cardinality among the business process, and have globally accepted standards, yet be efficient needs to be developed.

Figure 4.13 shows four different forms of conditional statements. Part (a) of the figure shows that in conventional client-server applications, the decision-making process is initiated by the client but the process execution is carried out at the server end. From Reference 21, we calculate that Figure 4.13(a) shows the conditional if-else statement with time complexity of $O(n^2)$, where n represents the number of conditional statements.

For every *if* statement, there is an action (*else*) statement that contains some executables in the form of resolution logic. Figure 4.13(b) shows that one if statement can contain the chain of many else statements. The exhaustive heuristic search uses the conditional clauses as shown in Figure 4.13(b) and contains the complexity of $O(M^N)$, where N are the

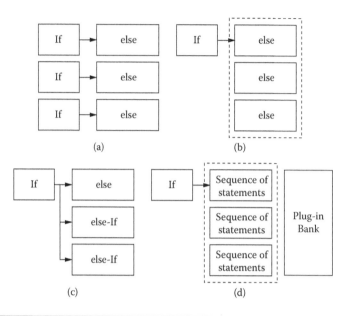

Figure 4.13 The proposed casual reasoning classification.

computations for finding the best match and M are the number of else statement possibilities. Figure 4.13(c) shows that the nested conditional statements carry complexity of $O(n^{2n})$ and are used in the circumstances of extreme conditional checks. This heuristics-based approach not only allows selecting the closest match of the rule but also allows us to statistically monitor the accuracy of the heuristic algorithms at work. We randomly assign weights to the device constraints (which we define in the normal functionality model, a device policy that is downloaded to the device at the initial configuration level) and to the constraints defined in the rule. Their effectiveness (score given by the user after rule execution on how much this solution has helped) and the rule's execution part (*else* part) count to calculate the value for a data structure called the affectivity factor (the use of the affectivity factor currently is not in the scope of this research).

An example of the operation is shown in Figure 4.14.

The device requirements/functional specifications along with the conditional constraints defined in the policy engine

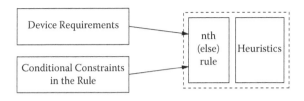

Figure 4.14   Hybrid intelligence generation model for rule generation.

generate the "right rule" along with the heuristics needed to generate it that help the system to grow in knowledge. In the scenario described in this research, when the abnormalities are sorted out, the policy engine generated rules help generate the healing policy.

References

1. Koscielny, J. and Syfert, M. Fuzzy logic application to diagnostics of industrial processes, Proceedings of Safeprocess 03, Washington, D.C., pp. 711–717.
2. Agrawal, D., Calo, S. B., Giles, J., Lee, K.-W., and Verma, D. C. Policy management for networked systems and applications, *Integrated Network Management*, 455–468, 2005.
3. Nayak, D. An adaptive and optimized security policy manager for wireless networks, Asia International Conference on Modeling and Simulation, 2007, pp. 155–158.
4. von Halle, B. *Business Rules Applied*, Wiley, New York, 2001.
5. Taveter, K. and Wagner, G. Agent-oriented enterprise modeling based on business rules? Proceedings of 20th Intl. Conf. on Conceptual Modeling (ER2001), LNCS, November 2001, Yokohama, Japan.
6. Forgy, C. RETE: a fast algorithm for the many pattern/many object pattern match problem, *Artificial Intelligence*, 19(1):17–37, 1982.
7. Flegkas, P., Trimintzios, P., Pavlou, G., and Liotta, A. Design and implementation of a policy-based resource management architecture, *Integrated Network Management*, 215–229, 2003.

8. Muruganantha, N. and Lutfiyya, H. Policy specification and architecture for quality of service management, *Integrated Network Management*, 535–548, 2003.

9. Kim, J. A., Taek, J. Y., and Hwang, S.-M. Rule-based component development, *SERA*, 70–74, 2005.

10. Wright, S., Chadha, R., and Lapiotis, G. (Eds.) Special issue on policy based networking, *IEEE Network*, 16, 2002.

11. Uszok, A., Bradshaw, J.M., Jeffers, R., Suri, N., Hayes, P. J., Breedy, M. R., Bunch, L., Johnson, M., Kulkarni, S., and Lott, J. KAoS policy and domain services: toward a description-logic approach to policy representation, deconfliction, and enforcement, *Policy*, 93, 2003.

12. Kagal, L., Finin, T. W., and Joshi, A. A policy language for a pervasive computing environment, *Policy*, 63, 2003.

13. Damianou, N., Dulay, N., Lupu, E., and Sloman, M. The Ponder policy specification language, *Policy*, 18–38, 2001.

14. Al-Shaer, E. S. and Hamed, H. H. Firewall policy advisor for anomaly discovery and rule editing, *Integrated Network Management*, 17–30, 2003.

15. Shankar, C., Ranganathan, A., and Campbell, R. An ECA-policy-based framework for managing ubiquitous computing environments, Mobiquitous 2005, San Diego, CA.

16. Bosch, J. Superimposition: a component adaptation technique, *Information and Software Technology*, 41(5):257–273, 1999.

17. Sztipanovits, J. and Karsai, G. Model-integrated computing, *IEEE Computer*, 30:110–112, 1997.

18. Balasubramanian, K., Gokhale, A., Karsai, G., Sztipanovits, J., and Neema, S. Developing applications using model-driven design environments, *IEEE Computer*, 33(2):33–40, 2006.

19. Karsai, G., Sztipanovits, J., Lédeczi, A., and Bapty, T. Model-integrated development of embedded software, *Proceedings of the IEEE*, 91(1):145–164, 2003.

20. The Business Rules Group. Defining Business Rules—What Are They Really? http://www.businessrulesgroup.org/first paper/BRG-whatisBR_3ed.pdf, July 2013.

21. Knuth, D. Big Omicron and big Omega and big Theta, *ACM SIGACT News*, 8(2), 1976.

Chapter 5

Related Work

Overview

This chapter surveys relevant research in distributed ubiquitous self-management systems and fault management solutions. It covers the salient features of pertinent systems and evaluates their suitability. The chapter concludes with an analysis of the shortcomings of current systems to motivate the need for a new bottom-up design of fault management reasoning facility and extensible self-management system to address the requirements for network management in a ubiquitous system.

Autonomic Network Management Systems

In this section, we compare the architecture proposed in this chapter with the solutions proposed in literature. The solutions reported in References 1 through 6 have scope in wireless sensor networks, References 7 through 13 deal with software development process, References 14 and 15 are hardware solutions, References 16 through 19 deal with pervasive computing

environments, and References 20 and 21 deal with grid infrastructure and RDBMS, respectively. There is not a lot of published work reported in the area of network management of
hybrid networks. Considering the unique nature, vast complexity, scale, and potential of u-Zone networks, it is essential
to propose an effective network management solution.

HYWINMARC

Shafique et al.[6] introduced a paradigm for network management in a hybrid network called HYbrid WIreless
Network Management ARChitecture (HYWINMARC). The
HYWINMARC is the first network management architecture
proposed for u-Zone networks that uses cluster heads to
manage the clusters through Simple Service Location Protocol
(SSLP). It collects the mobile device information through
Simple Network Management Protocol (SNMP). The cluster
header chooses the suitable mobile code agent and the mobile
code execution environment (MCEE) hosts mobile agents for
device management. Both AHSEN and HYWINMARC split
their activity and architectural completeness, consideration of
hybrid nature of clients, specification of executable components, and HYWINMARC fails to answer the considerations
raised in the previous section of this chapter. The selection of
executable components, consideration that "one device's management solution can be fatal for another," the wisdom of use
and architecture of mobile agents, post-management confirma-
tion that the problem is resolved, and self-growing nature are
some of the features that are not present in HYWINMARC.
More feature comparisons can be found in Reference 23.

AMUSE

The Agent-based Middleware for Context-aware Ubiquitous
Services (AMUSE)[24] is multi-agent-based middleware for ubiquitous computing environments. The design goal of AMUSE

is service construction for QoS-aware service provisioning considering the multiple contexts. AMUSE uses peer-to-peer communication among mobile nodes to enforce a cluster level self-management called "self-managed cluster." The rule-based profile assignment (called entity agentification in Reference 24) and Inter-Agent Relationship (IAR) define the link of one entity with another in a network. Now, using this knowledge, it becomes easier to detect the exact location of the problem in a network. After the entity agentification process, various entities are composed into applications. This publish-subscribe service can create serious issues like service consistency, synchronization, and coordination as discussed in Reference 25. Although AMUSE gives a more distinct hierarchy for the management framework to define the boundaries and performance optics, the payload attached with agents may not work for weaker nodes. This can be a big drawback in a heterogeneous environment. The monitoring in AMUSE is fault-based; that is, when the fault is recorded only then is it reported, whereas in AHSEN we use hybrid monitoring that is both fault-based and periodic.

RoSES

Morikawa[26] proposes that there are certain faults that cannot be removed through configuration of the system, which means that RoSES,[27] which solves all the faults through reconfiguration only, does not fulfill Morikawa's definition of self-management. It assigns state variables to different components of the system and monitors their states over time. The change of state of a variable (state variable staleness) triggers reconfiguration of the whole component to its original form and thus the whole process that was in the pipeline has to be performed again. RoSES uses both hot and cold swapping to replace components. However, it does not consider the areas like mobile code management and management function classification (i.e., consider self-* functions other than self-configuration), it contains limited system knowledge but has fast comeback time, and it is good for systems with closed-end specifications.

AMUN

The AMUN[28] is an autonomic middleware that deals with intra-cluster communication issues better than RoSES[27] does with higher support for multi-application environments. Both architectures rely mainly on regressive configuration and do not address issues such as higher traffic load leading to management framework failure, link level management, and framework synchronization. It keeps track of all the network traffic and relies in even on the traffic logs to dispatcher for fault notification. It also has a configuration-based recovery system and gives transport layer fault management.

SMHMS

Anerousis et al.[29] present Self Managing Health Monitoring Systems (SMHMS), a health monitoring and control system for application server environments. The scope of their work is thus narrower than that of AHSEN. The paper presents a model in which class-specific healing policies are assigned to application servers. Four classes are reported, which relate to memory usage, average response time, server usage, and amount of work performed. The policies used are threshold based, that is, when a class threshold is exceeded a restart (rejuvenation) action is triggered. To monitor the four preselected health conditions of interest, the existing middleware publish-subscribe reporting mechanism interface is employed. This is in contrast with AHSEN, which introduces a hybrid of "need-based/periodic" monitoring into the node via the NFM-assisted system variable matching. Although the idea of monitoring the selected parameters is quite the same in both AHSEN and the architecture proposed in Reference 29, the functionality has marked difference in terms of effectiveness, context management, and network payload. Moreover, Anerousis et al. have similar reservations with publish-subscribe mechanism to the AMUSE.[24] The AHSEN contains a multi-staged policy failure handling mechanism, but SMHMS triggers action upon class threshold rejuvenation. A generalized survey of autonomic systems can be found in Reference 20.

Case-Based Reasoning Systems

CHEMREG

CHEMREG[30] is a large knowledge-based system used by Air Products and Chemicals, Inc. to support compliance with regulatory requirements for communicating health and safety information in the shipping and handling of chemical products. CHEMREG concentrated on one of the knowledge bases in this system for the case-based reasoner. It generates estimates of hazard data from similar products using an existing product database as its case library. Although some refinements remain, the performance of the case-based reasoner has met its design goals.

JColibri

JColibri[31] is an object-oriented framework in Java for building CBR systems. JColibri is a software artifact that promotes software reuse for building CBR systems, integrating the application of well-proven software engineering techniques with a knowledge level description that separates the problem solving method that defines the reasoning process from the domain model that describes the domain knowledge. Framework instantiation is supported by a graphical interface that guides the configuration of a particular CBR system, alleviating the steep learning curve typical for these systems.

IBROW Project

The IBROW project intends to provide an Internet-based brokering service for reusing problem-solving methods. The Unified Problem-Solving Method Description Language (UPML)[32] has been developed to describe and implement such architectures and components to facilitate their semi-automatic reuse and adaptation. The main drawback of UPML is that it is a highly sophisticated formal language whose complexity is not justified by supporting reasoning tools.

CBR*Tools

In the CBR community, the work by Michel Jaczynski[33,34] is closely related to the AHSEN presented here. CBR*Tools is an object-oriented framework, implemented in Java, designed to facilitate the development of CBR applications. They identify the following axes of variability: the delegation of reasoning steps, the separation of case storage and case indexing, the design of indexes as reusable components, and the design of adaptation patterns. The framework concentrates mainly on indexing, providing a large number of indexing alternatives.

NaCoDAW

NaCoDAW[34] is a conversational case-based reasoning environment that can be used to develop tools that address Navy and DoD decision aids tasks. Written in Java, it is a platform independent system that can run on any system containing a Java Virtual Machine (JVM). It has features like case libraries, browsing cases, problem-solving sessions, parameter settings, history viewing, library revision, etc.

Policy Engine

The Business Rules Group[35] defines a business rule as "a statement that defines and constrains some aspects of business." It is intended to assert business structure or to control or influence the behavior of the business. The Object Management Group (OMG) is working on business rules semantics.[36] Nevertheless, several classifications of different rule types have emerged.[35,37,38]

RETE

In Reference 37, business rules are classified into four different types, such as integrity rules, derivation rules, reaction rules, and demonic assignments. A well-known algorithm for matching rule conditions is RETE.[39] For business rule expression, rule markup language is needed. Currently, BRML (Business Rule Markup Language),[40] Rule Markup Language (RuleML),[41] and Semantic Web Rule Language (SWRL)[42] are proposed as rule markup languages.

Business Rule Markup Language (BRML)

IBM took the initiative of developing BRML for the Electronic Commerce Project.[40] BRML is an XML encoding, which represents a broad subset of KIF. The Simple Rule Markup Language (SRML)[43] is a generic rule language consisting of a subset of language constructs common to the popular forward-chaining policy engines. Another rule markup approach is SWRL, a member submission to the W3C. It is a combination of OWL DL and OWL Lite sublanguages of the OWL Web Ontology language.[42] SWRL includes an abstract syntax for Horn-like rules in both of its sublanguages.

JSR-94 (Java Specification Request)

Most recently, the Java Community Process finished the final version of their Java Policy Engine API. The JSR-94 (Java Specification Request) was developed in November 2000 to define a runtime API for different policy engines for the Java platform. The API prescribes a set of fundamental policy engine operations based on the assumption that clients need to be able to execute a basic multiple-step policy engine cycle (parsing the rules, adding objects to an engine, firing rules, and getting the results).[44] It does not describe the content representation of the rules. The Java Rule API is already supported (at least partially) by a number of policy engine vendors (cf. Drools,[41] ILOG,[43] or JESS[45] to support interoperability.

Similarity Approaches

ELIXIR

An early approach for XML approximate retrieval is ELIXIR[46] that allows approximate matching in data content (tree leaves) and ranks the results according to the matching degree, disregarding structure in the evaluation. No structural heterogeneity is considered, and vocabulary heterogeneity is allowed only for data content elements, not for tags.

XIRQL and XXL

More sophisticated approaches, like XIRQL[47] and XXL[48] accept approximate matching at nodes and then discuss how to combine weights depending on the structure. Vocabulary heterogeneity is supported for content and element tags, but no structural heterogeneity is allowed. Conditions on document structure are interpreted as filters; thus, they need to be exactly satisfied. XXL supports a similarity operator and, to use this operator, the user should be aware of the occurrence of similar keywords or element tags.

Fuzzy Weights

In the approach proposed by Damiani and Tanca[49] both XML documents and queries are modeled as graphs labeled with fuzzy weights capturing the information relative relevance. They propose to employ both structure related weighting (weight on an edge) and tag related weighting (weight on a node). Some criteria for weighting are proposed such as the weight decreases as you move away from the root, and the weight depends on the dimension of the sub-tree. Shortcut edges are considered, thus allowing the insertion of nodes, which weight is a function of weights of edges. The match score is a normalized sum of weight of edges.

Relaxed Weights

In the tree relaxation approach,[50] exact and relaxed weights are associated with query nodes and edges. The score of a match is computed as the sum of the corresponding weights, and the relaxed weight is a function of the transformations applied to the tree. The considered transformations are relax node, replacing the node content with a more general concept; delete node, making a leaf node optional by removing the node and the edge linking it to its parent; relax edge, transforming a parent/child relationship to an ancestor/descendant relationship; and promote node, moving a node up in the tree structure (and the corresponding sub-tree).

ApproXQL

The approXQL[51] approach can also handle partial structural matching. However, all the paths in the query are required to occur in the document. The allowed edit operations on the document tree are delete node, insert intermediate node, and relabel node. The score of match is a function of the number of transformations, each one of which is assigned with a user-specified cost.

Query Decomposition Approach

Amer-Yahia et al.[52] account for both vocabulary and structural heterogeneity and propose scoring methods that are inspired by tf*idf and rank query answers on both structure and content. Specifically, twig scoring accounts for all structural and content conditions in the query. Path scoring, by contrast, is an approximation of twig scoring that loosens the correlations between query nodes when computing scores. The key idea in path scoring is to decompose the twig query into paths, independently compute the score of each path, and then combine these scores.

Our approach allows for more significant structural variations and can be considered a step forward in the treatment of structural heterogeneity in the context of XML. All the considered approaches, indeed, enforce at least the ancestor–descendant relationship in pattern retrieval, whereas our approach also allows inverting and relaxing this relationship. Moreover, our approach is highly flexible because it allows choosing the most appropriate structural similarity measures according to the application semantics.

Fault Detection

Expert Systems in Fault Detection

Relevant recent research work is reported in References 53 through 62. In these knowledge-driven techniques, although the governing elements are symbolic, numeric computations still play an important role in providing certain kinds of information for making decisions. Various methodologies have been proposed for the combination of knowledge-based techniques with numerical techniques. Frank[63,64] and Patton[65] consider that the combination of both approaches in an effective way offers an appropriate solution for most situations.

Neural Networks in Fault Detection

Research work on neural networks in on-line fault detection processes includes References 66 through 86. Scientists have pointed out drawbacks of neural networks especially of the back propagation networks that make them undesirable for on-line fault diagnosis applications. One of their limitations in the on-line fault detection process is the high accuracy of the measurements needed in order to calculate the evolution of faults. Fault detection usually makes use of measurements taken by instruments that may not be sensitive enough or that may produce noisy data. In this case, the neural network may not be successful in identifying faults. It is nearly always necessary to pre-process the data so that only meaningful parameters are presented to the net.

Qualitative Simulation in On-Line Fault Detection

Research work on this method includes the qualitative reasoning of De Kleer and Brown,[87] the qualitative process theory of Collins and Forbus,[88] the qualitative simulation of Dvorak and Kuipers,[89] and a lot of research work in diagnosis.[90–98] The main advantage of this approach is that accurate numerical knowledge and time-consuming mathematical models are not needed. On the other hand, this method only offers solutions in cases where high numerical accuracy is not needed.

On-Line Expert Systems in Fault Detection

On-line diagnostic systems emerging from recent research areas usually combine quantitative methods of fault detection with qualitative methods. This combination allows the evaluation of all available information and knowledge about the system for fault detection.

One of the main characteristics of this system is that, in parallel to the knowledge base of the traditional expert system, a database exists with information about the present state of the process. This information is derived on-line from the sensors. The database is in a state of continuous change. The knowledge base of the system contains both analytical knowledge and heuristic knowledge about the process. The knowledge engineering task comprises different knowledge sources and structures. The inference engine combines heuristic reasoning with algorithmic operation in order to reach a specific conclusion.

Angeli and Atherton[99] have developed an on-line expert system to detect faults in electro-hydraulic systems using on-line connections with sensors, signal analysis methods, model-based strategies, and deep reasoning techniques. Expert knowledge is contained primarily in a model of the expert domain. The final diagnostic conclusions will be conducted after interaction among various sources of information. Koscielny and Syfert[100] presented the main problems that appear in diagnostics of large-scale processes in the chemical, petrochemical, pharmaceutical, and power industries, and propose an algorithm for decomposition of a diagnostic system, dynamical creation of fault isolation threads, and multiple fault isolation assuming single fault scenarios.

References

1. Boonma, P. and Suzuki, J. BiSNET: A biologically-inspired middleware architecture for self-managing wireless sensor networks, *Computer Networks*, 51(16):4599–4616, 2007.
2. Trumler, W., Ehrig, J., Pietzowski, A., Satzger, B., and Ungerer, T. A distributed self-healing data store, *ATC*, 458–467, 2007.

3. Trumler, W., Helbig, M., Pietzowski, A., Satzger, B., and Ungerer, T. Self-configuration and self-healing in AUTOSAR, 14th Asia Pacific Automotive Engineering Conference (APAC-14), Hollywood, CA, August 5–8, 2007.
4. Bokareva, T., Bulusu, N., and Jha, S. SASHA: towards a self-healing hybrid sensor network architecture, Proceedings of the 2nd IEEE International Workshop on Embedded Networked Sensors (EmNetS-II), Sydney, Australia, May 2005.
5. Liu, P. ITDB: an attack self-healing database system prototype, *DISCEX*, 2:131–133, 2003.
6. Sajjad, A., Jameel, H., Kalim, U., Han, S. M., Lee, Y.-K., and Lee, S. AutoMAGI—an autonomic middleware for enabling mobile access to grid infrastructure, Joint International Conference on Autonomic and Autonomous Systems and International Conference on Networking and Services (icas-icns'05), p. 73, 2005.
7. Krena, B., Letko, Z., Tzoref, R., Ur, S., and Vojnar, T. Healing data races on-the-fly, Proceedings of the 2007 Workshop on Parallel and Distributed Systems: Testing and Debugging (PADTAD 2007), London, England, July 9, 2007.
8. Shehory, O. A self-healing approach to designing and deploying complex, distributed and concurrent software systems, *Lecture Notes in AI*, Vol. 4411, Bordini, R., Dastani, M., Dix, J., and El Fallah Seghrouchni, A. (Eds.), Springer Verlag, pp. 3–11, 2006.
9. Breitgand, D., Goldstein, M., Henis, E., Shehory, O., and Weinsberg, Y. PANACEA: towards a self-healing development framework, *Integrated Network Management,* 169–178, 2007.
10. Siewert, S. and Pfeffer, Z. An embedded real-time autonomic architecture, IEEE DenverTechnical Conference, April 2005.
11. Can, W., Yang, L., and Jianjun, B. A biological formal architecture of self-healing system, *SMC*, 6: 5537–5541, 2004.
12. Shen, C., Pesch, D., and Irvine, J. A. Framework for self-management of hybrid wireless networks using autonomic computing principles, Proceedings of the 3rd Annual Communication Networks and Services Research Conference (Cnsr'05), May 16–18, 2005. IEEE Computer Society, Washington, D.C., pp. 261–266.
13. Grishikashvili, E., Pereira, R., and Taleb-Bendiab, A. Performance evaluation for self-healing distributed services, Proceedings of the 11th International Conference on Parallel

and Distributed Systems, Workshops (Icpads'05), July 20–22, 2005. IEEE Computer Society, Washington, D.C., pp. 135–139.

14. Zhang, X., Dragffy, G., Pipe, A. G., Gunton, N., and Zhu, Q. M. A reconfigurable self-healing embryonic cell architecture, *Engineering of Reconfigurable Systems and Algorithms*, 134–140, 2003.

15. Wile, D. and Egyed, A. An externalized infrastructure for self-healing systems, Proceedings of the 4th Working IEEE/IFIP Conference on Software Architecture (WICSA), Olso, Norway, pp. 285–288, June 2004.

16. Trumler, W., Petzold, J., Bagci, F., and Ungerer, T. AMUN: an autonomic middleware for the Smart Doorplate Project, *Personal Ubiquitous Computing*, 10(1):7–11, 2005.

17. Wang, F. and Li, F.-Z. The design of an autonomic computing model and the algorithm for decision-making, *GrC*, 270–273, 2005.

18. Gangadhar, D.K. Meta dynamic states for self healing autonomic computing systems, 2005 IEEE International Conference on Systems, Man and Cybernetics, 1:39–46, 2005.

19. Zenmyo, T., Yoshida, H., and Kimura, T. A self-healing technique using reusable component-level operation knowledge, *Cluster Computing*, 10(4):385–394, 2007.

20. Caseau, Y. Self-adaptive and self-healing message passing strategies for process-oriented integration infrastructures, 11th IEEE International Conference and Workshop on the Engineering of Computer-Based Systems (ECBS'04), p. 506, 2004.

21. Sharmin, M., Ahmed, S., and Ahamed, S. I. MARKS (middleware adaptability for resource discovery, knowledge usability and self-healing) for mobile devices of pervasive computing environments, Proceedings of the Third International Conference on Information Technology: New Generations (ITNG 2006), Las Vegas, NV, pp. 306–313, April 2006.

22. Chaudhry, S. A., Akbar, A. H., Kim, K.-H., Hong, S.-K., and Yoon, W.-S. *HYWINMARC: An Autonomic Management Architecture for Hybrid Wireless Networks*, Network Centric Ubiquitous Systems, 2006.

23. Chaudhry, J. A. and Park, S. AHSEN—Autonomic Healing-based Self-management Engine for Network management in hybrid networks, The Second International Conference on Grid and Pervasive Computing (GPC07), 2007.

24. Takahashi, H., Suganuma, T., and Shiratori, N. AMUSE: an agent-based middleware for context-aware ubiquitous services, *ICPADS,* 1:743–749, 2005.

25. Lee, Y., Chaudhry, J. A., Min, D., Han, S., and Park, S. A dynamically adjustable policy engine for agile business computing environments, *Lecture Notes in Computer Science, Advances in Data and Web Management,* Joint 9th Asia-Pacific Web Conference (APWeb/WAIM 2007), 785–796.

26. Morikawa, H. The design and implementation of context-aware services, Proceedings of IEEE SAINT-w, pp. 293–298, 2004.

27. Shelton, C. and Koopman, P. Improving system dependability with alternative functionality, DSN04, June 2004.

28. Trumler, W., Petzold, J., Bagci, F., and Ungerer, T. AMUN–Autonomic Middleware for Ubiquitous eNvironments Applied to the Smart Doorplate Project, International Conference on Autonomic Computing (ICAC-04), New York, May 17–18, 2004.

29. Anerousis, N., Black, A., Hanson, S., Mummert, L., and Pacifi, G. Health monitoring and control for application server environments, *Integrated Management* (IM'05), 2005.

30. Wilson, K. D. CHEMREG: using case-based reasoning to support health and safety compliance in the chemical industry, *AI Magazine,* 19(1): 1998.

31. Bello-Tomás, J. J., González-Calero, P. A., Díaz-Agudo, B., and Colibri, J. An object-oriented framework for building CBR systems, *ECCBR,* 32–46, 2004.

32. Fensel, D., Motta, E., van Harmelen, F., Benjamins, V. R., Crubezy, M., Decker, S., Gaspari, M., Groenboom, R., Grosso, W., Musen, M., Plaza, E., Schreiber, G., Studer, R., and Wielinga, B. The unified problem-solving method development language upml, *Knowledge and Information Systems,* 5(1):83–131, February 2003.

33. Jaczynski, M. Modèle et plate-forme à objets pour l'indexation des cas par situations comportementales: application à l'assistance à la navigation sur le Web. PhD thesis, L'Université de Nice-Sophia Antipolis, 1998.

34. Jaczynski, M. and Trousse, B. An object-oriented framework for the design and the implementation of case-based reasoners, Proceedings of the 6th German Workshop on Case-Based Reasoning, 1998.

35. The Business Rules Group. Defining Business Rules—What Are They Really? http://www.businessrulesgroup.org/first paper/BRG-whatisBR_3ed.pdf, July 2013.
36. von Halle, B. *Business Rules Applied*, Wiley, New York, 2001.
37. Taveter, K. and Wagner, G. Agent-oriented enterprise modeling based on business rules? Proceedings of 20th Intl. Conf. on Conceptual Modeling (ER2001), LNCS, November 2001, Yokohama, Japan.
38. Russell, S. and Norvig, P. *Artificial Intelligence–A Modern Approach*, 2nd ed., Prentice Hall, Englewood Cliffs, NJ, 2003.
39. Forgy, C. RETE: a fast algorithm for the many pattern/many object pattern match problem, *Artificial Intelligence*, 19(1):17–37, 1982.
40. IBM T.J. Watson Research. Business Rules for Electronic Commerce Project, http://www.research.ibm.com/rules/home.html, 1999.
41. RuleML Initiative. http://ruleml.org/.
42. W3C. OWL Web Ontology Language Overview, http://www.w3.org/TR/owl-features/, W3C Recommendation 10, February 2004.
43. ILOG. Simple Rule Markup Language (SRML), http://xml.coverpages.org/srml.html, 2001.
44. Chaudhry, J. A. and Park, S. Using artificial immune systems for self-healing in hybrid networks, in *Encyclopedia of Multimedia Technology and Networking*, Idea Group Inc., http://www.igi-global.com/about/, 2006.
45. JESS. Java Policy Engine, http://herzberg.ca.sandia.gov/jess.
46. Chinenyanga, T.T. and Kushmerick, N. An expressive and efficient language for XML information retrieval, *Journal of the American Society for Information Science and Technology*, 53:438–453, 2002.
47. Fuhr, N. and Grossjohann, K. XIRQL: a query language for information retrieval in XML documents, Proceedings of the 24th ACM SIGIR Conference on Research and Development in Information Retrieval, pp. 172–180, 2001.
48. Theobald, A. and Weikum, G. Adding relevance to XML, Proceedings of the Third International Workshop on the Web and Databases, Suciu, D. and Vossen, G. (Eds.), LNCS, pp. 105–124, 1997.

49. Damiani, E. and Tanca, L. Blind queries to XML data, Proceedings of the 11th International Conference on Database and Expert Systems Applications, pp. 345–356, 2000.

50. Kanza, Y. and Sagiv, Y. Flexible queries over semistructured data, Proceedings of the 20th ACM Symposium on Principles of Database Systems, 2001.

51. Schlieder, T. and Naumann, F. Approximate tree embedding for querying XML data, Proceedings of the ACM SIGIR Workshop on XML and Information Retrieval, 2000.

52. Amer-Yahia, S., Koudas, N., Marian, A., Srivastava, D., and Toman, D. Structure and content scoring for XML, Proceedings of the 31st International Conference on Very Large Data Bases, pp. 361–372, 2005.

53. Forbus, K. D. and Falkenhainer, B. Self-explanatory simulations: an integration of qualitative and quantitative knowledge, *IJCAI*, 380–387.

54. Oyeleye, O., Finch, F., and Kramer, M. A robust event-oriented methodology for diagnosis of dynamic process systems, *Computers and Chemical Engineering*, 14:1379–1398, 1990.

55. Dvorak, D. and Kuipers, B. Process monitoring and diagnosis, *IEEE Expert*, 6(4), 1991.

56. Yu, C. and Lee, C. Fault diagnosis based on qualitative/quantitative process knowledge, *AIChE Journal*, 37:617–628, 1991.

57. Benouarets, M. and Dexter, A. On-line fault detection and diagnosis using fuzzy models, in IFAC Workshop, On-line Fault Detection and Supervision in the Chemical Process Industries, Newcastle, pp. 213–218, June 12–13, 1995.

58. Kay, H. Robust identification using semiquantitative methods, IFAC Symposium, Fault Detection, Supervision and Safety for Technical Processes, Kingston Upon Hull, U.K., pp. 277–282, August 26–28, 1997.

59. Milne, R., Nicol, C., Travé-Massuyès, L., and Quevedo, J., Model based aspects of the TIGER gas turbine condition monitoring system, IFAC Symposium, Fault Detection, Supervision and Safety for Technical Processes, Kingston Upon Hull, U.K., pp. 405–410, August 26–28, 1997.

60. Manders, E. and Biswas, G. FDI of abrupt faults with combined statistical detection and estimation and qualitative fault isolation, Proceedings Safeprocess 2003, Washington, D.C., pp. 339–344, 2003.

61. Nyberg, M. and Krysander, M. Combining AI, FDI and statistical hypothesis-testing in a framework for diagnosis, Proceedings Safeprocess 03, Washington, D.C., pp. 813–818, 2003.
62. Saludes, S., Corrales, A., Miguel, L., and Peran, J. A SOM and expert system based scheme for fault detection and isolation in a hydroelectric power station, Proceedings Safeprocess 03, Washington, D.C., pp. 999–1004, 2003.
63. Frank, P. Fault diagnosis in dynamic systems using analytical and knowledge based redundancy—a survey and some new results, *Automatica*, 26(3):459–474, 1990.
64. Frank, P.M. Analytical and qualitative model-based fault diagnosis: A survey and some new results, *European Journal of Control*, 2:6–28, 1996.
65. Patton, R.J., Frank, P. M., and Clark, R. N. *Fault Diagnosis in Dynamic Systems, Theory and Application*, Prentice Hall, Englewood Cliffs, NJ, 1999.
66. Himmelblau, D. M. Fault detection and diagnosis—today and tomorrow, Proc. of the 1st IFAC Workshop on Fault Detection and Safety in Chemical Plants, Kyoto, pp. 95–105, Sept. 28–Oct. 1, 1986.
67. Kramer, M. A. and Leonard, J. A. Diagnosis using backpropagation neural networks; analysis and criticism, *Computer and Chemical Engineering*, 14:1323–1338, 1997.
68. Rengaswamy, R. and Venkatasubramanian, V. An integrated framework for process monitoring, diagnosis, and control using knowledge-based systems and neural networks, in IFAC Symposium, On-line fault detection and supervision in the chemical process industries, Newark, DE, pp. 49–55, April 22–24, 2003.
69. McDuff, R. J. and Simpson, P. K. An adaptive resonance diagnostics system, *Journal of Neural Network Computing*, 2:19–29, 1990.
70. Foss, B. A. and Johansen, T. A. An integrated approach to on-line fault detection and diagnosis including artificial neural networks with local basis functions, IFAC Symposium in On-Line Fault Detection and Supervision in the Chemical Process Industries, Newark, DE, pp. 207–213, April 22–24, 2003.
71. Zhang, J. and Morris, A. On-line process fault diagnosis using fuzzy neural networks, *Intelligent Systems Engineering*, 3:37–47, 1994.

72. Yu, D., Gomm, J., and Williams, D. Diagnosing sensor faults in a chemical process via RBF network, IFAC Symposium, Fault Detection, Supervision and Safety for Technical Processes, Kingston Upon Hull, U.K., pp. 893–898, August 26–28, 1997.

73. Akhmetov, D. and Dote, Y. Novel "on-line" identification procedure using artificial neural network, IFAC Workshop, On-line Fault Detection and Supervision in the Chemical Process Industries, Newcastle, U.K., June 12–13, 1995.

74. Hessel, G., Schmitt, W., and Weiss, F.-P. A neural network approach for acoustic leak monitoring at pressurized plants with complicated topology, IFAC Workshop, On-line Fault Detection and Supervision in the Chemical Process Industries, Newcastle, U.K., pp. 83–88, June 12–13, 1995.

75. Fujiwara, T., Tsushi, T., and Nishitani, H. Failure detection by auto-associative neural networks, IFAC Workshop, On-line Fault Detection and Supervision in the Chemical Process Industries, Newcastle, U.K., pp. 33–38, June 12–13, 1995.

76. Alessandri, A. and Parisini, T. Direct model-based fault diagnosis using neural filters, In IFAC Symposium, Fault Detection, Supervision and Safety for Technical Processes, Kingston Upon Hull, U.K., pp. 343–348, August 26–28, 1997.

77. Benkhedda, H. and Patton, R. Information fusion in fault diagnosis based on BSpline network, IFAC Symposium, Fault Detection, Supervision and Safety for Technical Processes, Kingston Upon Hull, U.K., pp. 669–674, August 26–28, 1997.

78. Schubert, M., Koppen-Seliger, B., and Frank, P. Recurrent neural networks for nonlinear system modelling in fault detection, IFAC Symposium, Fault Detection, Supervision and Safety for Technical Processes, Kingston Upon Hull, U.K., pp. 701–706, August 26–28, 1997.

79. Yu, D., Gomm, J., and Williams, D. Diagnosing sensor faults in a chemical process via RBF network, IFAC Symposium, Fault Detection, Supervision and Safety for Technical Processes, Kingston Upon Hull, U.K., pp. 893–898, August 26–28, 1997.

80. Shayler, P., Goodman, M., and Ma, T. Applications of neural networks in automotive engine management systems, IFAC Symposium, Fault Detection, Supervision and Safety for Technical Processes, Kingston Upon Hull, U.K., pp. 899–906, August 26–28, 1997.

81. Yang, S., Chen, B., Wang, X., and McGreavy, C. Soft sensor based fault diagnosis of industrial fluid catalytic cracking reactor, IFAC Symposium, Fault Detection, Supervision and Safety for Technical Processes, Kingston Upon Hull, U.K., pp. 241–246, August 26–28, 1997.

82. Dimla, D., Lister, J., and Leighton, N. Sensor Fusion for cutting tool state identification in metal turning through the application of perceptron neural networks, IFAC Symposium, Fault Detection, Supervision and Safety for Technical Processes, Kingston Upon Hull, U.K., pp. 1143–1146, August 26–28, 1997.

83. Boudaoud, A. and Masson, M. On-line adaptive fuzzy diagnosis system: fusion and supervision, IFAC Symposium, Fault Detection, Supervision and Safety for Technical Processes, Kingston Upon Hull, U.K., pp. 1195–1200, August 26–28, 1997.

84. Papadimitropoulos, A., Rovithakis, G., and Parasini, T. Fault detection in mechanical systems with friction phenomena: an on-line neural approximation approach, Proceedings Safeprocess 03, Washington, D.C., pp. 705–711, 2003.

85. Rzepiejewski, P., Syfert, M., and Jegorov, S. On-line actuator diagnosis based on neural models and fuzzy resoning: the DAMADICS Benchmark Study, Proceedings Safeprocess 03, Washington, D.C., pp. 981–986, 2003.

86. Can, C. and Danai, K. Fault diagnosis with a model-based recurrent neural network, Proceedings Safeprocess 03, Washington, D.C., pp. 699–704, 2003.

87. De Kleer, J. and Brown, J. A qualitative physics based on confluences, *Artificial Intelligence*, 24:7–84, 1984.

88. Collins, J. and Forbus, K. Reasoning about fluids via molecular collections, in *Readings in Qualitative Reasoning About Physical Systems*, Weld, D. and Kleer, J. (Eds.), M. Kaufman, Inc., pp. 503–507, 1987.

89. Dvorak, D. and Kuipers, B. Process monitoring and diagnosis, *IEEE Expert*, 6(4), 1994.

90. De Kleer, J. and Kurien, J. Fundamentals of model-based diagnosis, Proceedings Safeprocess 03, Washington, D.C., pp. 25–36, 2003.

91. Herbert, R. and Williams, G. An initial evaluation of the detection and diagnosis of power plant faults using a deep knowledge representation of physical behaviour, *Expert Systems*, 4:90–99, 1987.

92. Zhang, J., Roberts, P., and Ellis, J. Fault diagnosis of a mixing process using deep qualitative knowledge representation of physical behaviour, *Journal of Intelligent and Robotic Systems*, 3:103–115, 1990.

93. Vinson, J. and Ungar, L. Fault detection and diagnosis using qualitative modelling and interpretation, IFAC Symposium, On-Line Fault Detection and Supervision in the Chemical Process Industries, Newark, DE, pp. 121–127, April 22–24, 1993.

94. Whiteley, J. and Davis, J. Qualitative interpretation of sensor patterns using a similarly based approach, IFAC Symposium, On-Line Fault Detection and Supervision in the Chemical Process Industries, Newark, DE, pp. 115–121, April 22–24, 1990.

95. Konstantinov, K. and Yoshida, T. A method for on-line reasoning about the time profiles of process variables, IFAC Symposium, On-Line Fault Detection and Supervision in the Chemical Process Industries, Newark, DE, pp. 133–139, April 22–24, 1990.

96. Coghill, G., Chantler, M., Shen, Q., and Leitch, R. Towards model switching for diagnosis of dynamic systems, IFAC Symposium, Fault Detection, Supervision and Safety for Technical Processes, Kingston Upon Hull, U.K., pp. 545–552, August 26–28, 1997.

97. Han, Z. and Frank, P. Physical parameter estimation based FDI with neural networks, IFAC Symposium, Fault Detection, Supervision and Safety for Technical Processes, Kingston Upon Hull, U.K., pp. 283–288, August 26–28, 1997.

98. De Kleer, J. and Kurien, J. Fundamentals of model-based diagnosis, Proceedings Safeprocess 03, Washington, D.C., pp. 25–36, 2003.

99. Angeli, C. and Atherton, D. P. A model based method for an on-line diagnostic knowledge-based system, *Expert Systems*, 18(3):150–158, 1990.

100. Koscielny, J. and Syfert, M. Fuzzy logic application to diagnostics of industrial processes, Proceedings Safeprocess 03, Washington, D.C., pp. 711–717, 2003.

Chapter 6

Implementation

Overview

The focus of this chapter is on the data structures and algorithms for the efficient identification of fragments and regions in the target. Each fragment is a set of nodes bound by the ancestor–descendant relationship in the target. A general purpose indexing structure along with an indexing structure depending on the pattern P is employed for improving the performances of our approach. Fragments are merged into regions only when the similarity between P and the generated region is greater than the similarity between P and each single fragment. Target sub-trees covered by a region should be evaluated only by accessing nodes in the regions and information contained in the auxiliary indexing structures.

In the remainder of the chapter, we first present the indexing structures. Then, we discuss the algorithms for the construction of a list of fragments and for the creation of regions starting from such a list.

Implementation Details

Similarity-Based Inverted Index and Pattern Index

Directly evaluating the pattern on the target is inefficient and introduces scalability issues in the approach due to the tag and structural heterogeneity of the collection. For these reasons, a similarity-based inverted index (SII) is proposed. This index is independent from the retrieval pattern and is composed by a traditional inverted index coupled with a name-similarity table, which specifies relationships among tags in the collection relying on the semantic functions discussed previously.

This index allows us to easily identify the collection nodes with similar tags according to different criteria. Since the number of labels that normally occur in a target is sensibly smaller than the number of elements, the size of the name-similarity table is often contained and it can fit in main memory.

The SII index is built as follows. Starting from the labels of a target T, a traditional inverted index is created; that is, each distinct label l offering in the target is associated with the list of vertices labeled by l, according to the preorder rank. Then, tags in the collection are grouped according to each of the tag-based similarity functions and each group is progressively numbered. Each tag is finally associated with the list of group identifiers to which it belongs. Table 6.1 reports this association by name of a table. For example, programmer and programmers belong to the same class according to similarity function, whereas coder, coders, and writer belong to the same class according to another similarity function. In this way, when nodes tagged l should be retrieved in the target, the rows in the name similarity table corresponding to l are extracted and, according to one or more similarity measures, all the similar tags are easily identified. In case l does not belong to the name-similarity table, one of the tag-based similarity functions can be applied to identify a similar tag in the table.

Given a pattern P, for every node v in P, all the occurrences of nodes u in the target tree such that label $(v) \cong p$ label (u) are retrieved from the SII index according to the functions in S, and organized level by level in a pattern index. The pattern index therefore depends on the pattern that should be evaluated on the target. The number of levels in the index depends on the levels in T at which vertices occur with labels similar to those in the pattern. For each level, vertices are ordered according to the preorder rank. Depending on the label similarity functions in S, different pattern indexes can be generated.

The construction of the pattern index requires identifying the entries in SII with similar tags to those of P. For each node of P, this operation requires (in the worst case) checking the distinct labels that tag each entry of SII. Once the entries are identified, the corresponding nodes are inserted in the pattern index. Since the nodes in each entry are ordered according to the preorder rank in the target, the construction of the pattern index is performed in linear time. Therefore, the number of operations for each tag in P is in the order of where K is the number of entries and M is the maximal number of nodes in each entry. The number of operations for the construction of the entire pattern index for a pattern P is thus. We remark that the use of the name-similarity table does not affect the complexity of the approach in the worst-case analysis but only in the average case.

Table 6.1 Create Fragments

```
Require: Pl,F,v,l
{Pl: Pattern index
 F: current fragment
 V: vertex in current fragment
 L: a level in the pattern index}
 SF = Ø
 If l≤ size(pl)∧ head(pl(l) is not null then
   While head (pl(l)) is not null and precedes
     root(F) do
```

(continued)

Table 6.1   Create Fragments (continued)

```
Create a new fragment F` = (V`F, Ø) such that
  V`F = {head(pl(1))}
SF = SF ∪ {F`} ∪ createFragments(Pl, F`,
head(pl(1)),1+1)
Remove head(pl(1))
end while
while head (Pl(1)) is not null and is a
descendent of (root(F)) do
Insert head(Pl)) in the vertex of F
if head (Pl(1)) precedes v then
insert (root(F), head (Pl(1)) in the edges of F
else
insert (v, head(Pl(1)) in the edges of F
end if
SF = SF ∪ createFragments (Pl, F, head(Pl(1)),
  1+1
Remove head (pl(1))
end while
SF = SF ∪ createFragments (Pl, F, v, 1+1)
end if
Ensure return SF
```

Once the index is generated, the fragments are generated through the recursive algorithm mentioned previously—CreateFragments—and in the algorithm following—CreateListOfFragments. The main advantage of these algorithms is the identification of the fragments through a single visit of the pattern index so that the complexity of fragments construction is kept linear.

Create List of Fragments and Regions

Table 6.2   Create List of Fragments

```
Require: Pl
{Pl: Pattern index
}
```

(continued)

Table 6.2   Create List of Fragments (continued)

```
SF = ∅
for l = 1 to size(Pl)
While head(Pl))is not null
Create a new fragment F = (VF, ∅) such that VF =
  {head(Pl))}
SF = SF ∪ {F} ∪ createFragments(Pl,F,head(Pl(l))
  l+1)
Remove head{Pl(l)}
end while
end for
Ensure return SF
```

Let PI be the pattern index for a pattern *P*. Algorithm 2—
CreateListOfFragments—correctly identifies all fragments in PI.
All the atomic operations in the algorithm (checking whether
a node precedes another one, adding a node to a fragment,
removing a node from PI) require a constant number of operations. The algorithm visits each vertex in the pattern index
only once by removing in each level the vertices already
included in a fragment. Therefore, its complexity is in O(N),
where N is the number of nodes in PI.

Two fragments should be merged in a single region when,
relying on the adopted similarity function, the similarity of the
pattern with the region is higher than the similarity with the
individual fragments.

Whenever a document in the target is quite big and the
number of fragments is high, the regions that should be
checked can grow exponentially. To avoid such a situation,
we exploit the following locality principle: merging fragments
together or merging fragments to regions makes sense only
when the fragments/regions are close. Indeed, as the size of
a region tends to be equal to the size of the document, the
similarity decreases.

In order to meet such locality principle, regions are
obtained by merging adjacent fragments. Operatively, two
adjacent fragments can be merged when their common

ancestor v is not the root of the target. If it is not the root of the target, their common ancestor v becomes the root of the region and the roots of the fragments become the direct children of v. We remark that the common ancestor of two fragments can be obtained easily by traversing the parent link included in the nodes of the pattern index.

Combining the locality principle and the approach for merging two adjacent fragments, the following algorithm—CreateListOfRegions—is obtained. The algorithm works on the list of fragments SF obtained from CreateListOfFragments that is ordered according to the preorder rank of the roots of the fragments it contains. Once a possible region Ri is obtained, by merging two adjacent fragments SF(i − 1) and SF(i), the similarity Sim(P,R) is compared with the maximal value between Sim(P,SF(i − 1)) and Sim (P,SF(i)). If Sim(P, Ri) is the highest, SF(i − 1) is removed from the list and SF(i) substituted with Ri. Otherwise, SF(i − 1) is kept alone and we try to merge SF(i) with its right adjacent fragment. The process ends when all fragments in the list have been checked.

Table 6.3   Create List of Regions

```
Require: SF, P
{SF = List of fragments
P: Pattern
}
for i = 2 to size(SF) do
Try to generate region Rᵢ by merging SF(i-1)
    and SF(i)
If Rᵢ has been generated then
If Sim(P,R) ≥ max{Sim(P, SF(i-1), Sim(P,SF(i))}
then
Remove SF(i-1)
Substitute SF(i) with Rᵢ
end if
end if
end for
Ensure return SF
```

Since the common ancestor between F1 and F2 is the root of the target, the two fragments cannot be merged. The same behavior is observed for fragments F2 and F3. Since the similarity of P with R is higher than its similarity with F3 and F4, R is kept and F3,F4 is removed. At the end of the process we have regions {F1,F2,R}.

The construction of regions is quite fast because the target should not be explicitly accessed. All the required information is contained in the inverted indexes. Moreover, thanks to our locality principle, the number of regions to check is proportional to the number of fragments. Finally, the regions obtained through our process do not present all the vertices occurring in the target but only those necessary for the computation of similarity. Vertices appearing in the region but not in the pattern are evaluated through the pre/post-order rank of each node. For example, let P be a pattern, K the number of distinct labels in the SII index, and M the maximal size of an entry in SII. Moreover, let N be the number of nodes in the pattern index PI. The number of operations for the creation of regions starting from a pattern is $\theta(\max\{(|P|.K.M),(N.|P|)\})$.

Chapter 7

Prototype

We have implemented a prototype using Java Net Beans Integrated Development Environment and XML extensions for these experiments. The software diagrams used in the experiment are as follows.

Parsing of XML Files

JDOM makes use of standard Java coding patterns. Where possible, it uses the Java new operator instead of complex factory patterns, making object manipulation easy, even for the user. We feed the source to the JDOM objects and the parsed data is stored in the main memory in a file.

DOM Object for Each File Created

The XML DOM (Document Object Model) defines a standard way for accessing and manipulating XML documents. The DOM presents an XML document as a tree structure, with elements, attributes, and text.

Construct Tree Target

After the creation of DOM objects, the source is fed to the JDOM and a DOM object is created to handle the tree extracted from the JDOM. A location is provided to the object to store the results from the parsing.

Construct Inverted Index

In information technology, an inverted index (also referred to as postings file or inverted file) is an index data structure storing a mapping from content, such as words or numbers, to its locations in a database file, or in a document or a set of documents, in this case allowing full text search. The inverted file may be the database file itself, rather than its index. It is the most popular data structure used in document retrieval systems.[1] The inverted index helps us tremendously in tree traversal (Figure 7.1).

Query Processing

A query is entered at this stage (in our case, the error report) and related fragments in the repository are searched.

Construct Pattern Index

An index is created because of the query and DOM tree match, which is called a pattern index. This structure is relatively smaller than the DOM structure and serves as a schema for later manipulations.

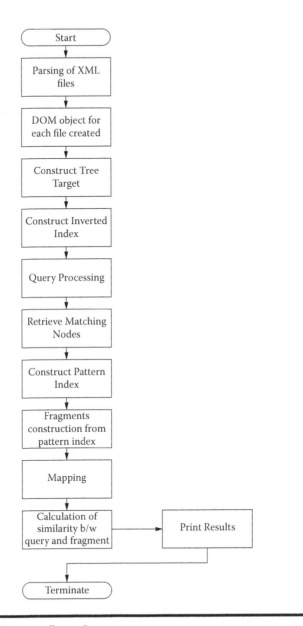

Figure 7.1   Process flow chart.

Fragments Construction Form Pattern Index

Depending on how relaxed the query is, various pattern fragments are gathered at this stage. We align the context fragments gathered from the fault context into an array data structure. The fragments into the transformed space are stringed together with the relevant fault context fragment.

The sequence diagram in Figure 7.2 shows the steps in creation of a transformed space and extraction of related fragments (Figure 7.3 and Figure 7.4).

TreeOperations: Main class where the XML parser class is invoked and the construction of tree with DB as the root node and each XML file as child sub-tree to DB. Then the

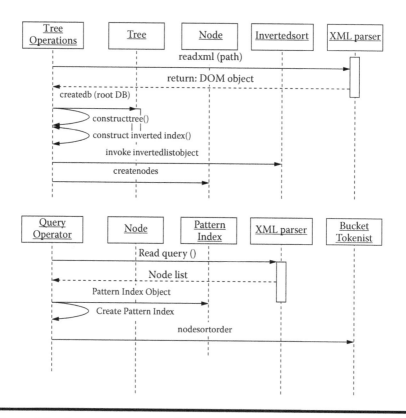

Figure 7.2   Process sequence diagram.

construction of invertedindex is done by traversing the tree and storing the invertedindex object in a file.

Tree class: It is a data structure containing the vector of children,Label,value of the label and the arraylist of five values.

Node class: Node contains string label and arraylist of five values.

Commonconstants class: It contains all the paths of XML collection and constant variables.

Invertedlist class: The object of this class contains the string label and the linked list of corresponding nodes to build the inverted index.

Xmlparser class: Parsing XML files and generating a DOM tree for each file.

QueryOperations class: This class takes the query construct the tree, opens the invertedindex file, and then retrieves the matching labels' corresponding nodes. Construction of patternindex is done by adding in an arraylist of nodes and sorting it according to level and preorder value.

Patternindex class: The object of this class contains a linked list of nodes and the distance d of arraylist.

Bucket class: This object stores the pattern tree node and corresponding matching similar nodes of a fragment tree in arraylist.

Bucketconstruct class: This class takes the fragment tree and pattern tree and constructs the bucket for each pattern tree node.

Similarity vector class: In the class, the distance vector for each node of fragments and pattern tree are computed.

Mapping class: In this, all possible mappings between pattern tree and fragment tree are made and corresponding similarity value is calculated. The maximum of all is given as the similarity value between fragment tree and pattern tree.

Similarity class: This class retrieves other similar labels that match the query before construction of the pattern index. Using the similarity table, the similar labels are known.

Tree

-Private Vector <Tree> Child
-Private String Data
-Private String Tent
-Private Array wst (integer> Alist
+getAlist <integer>()
+SetAlist (Arraylist<Integer>)()
+getchildren()
+gettext()
+settent()
+setpreorder()
+setpostorder()
+getpreorder()
+getpostorder()
+getpostorder(integeer)()

---

bucketcostinvert

-Arraylist <bucket> bucketcost
+Public double similar (pattern index, pattern index)()

XML Parser

-Public Element readsml (String Path) : string

Invertedlis

-Private string tent
-Private String Data
-PRivate Linkedlist <arraylist<integer>>alist
+Public llinkedlist <arraylist <integer> getalist()>()
+Public void setalist (linkedlist<array list <integer> list)()
+Public string getdata()()
+public void setdata()()
+public string gettent()()
+public void settent(string tent)()

Query Operation

-int prevalue
-int hight : <unspecified> = 0
-int nodeordertraversalvalue
-public linkedlist <Node> ll
-public hashset <string> queryt
-Public araylist <bucket> bucketlist
-public array <patterninded> patternindex
-public arraylist <patternindex> final patternindex
-public patternindex frequencytree
-public arraylist <[atternindex> fragmentslist
-public arraylist<invertedlist> inverted index
+Public void (string[] args)()
+public void intlistfrequency (tree Node)()
+public void constructtree(element root, tree node)()
+Public void preorder (tree node)()
+Public void postorder(Tree Node)()
+Public void setsiblings (arraylist<string> tent, Tree Node)()
+public void setlevel (tree Node, int height)()
+Public void setpostion (Tree node, int position)()
+public void generatepatternindex (arraylist <string> Querystring (s))()
+public void commonparent (node m, Node n, Tree Node)()
+Public boolean containnewfunc(node n, vector <tree> v)()

Tree Operations

-Prevalue : int
-Postvalue : int
-Height : int
-Preordertraversalvalue : long
-Arraylist(invertedlist> : long
-Foldername : string
-invertedfile : string
-Vector <tree> nodes
+Public main (string[] args)()
+Public Tree init(tree node, string path)()
+Public void initialize (Tree node)()
+Public void Preorder (Tree node)()
+Public void Postorder (Tree Node)()
+Public void setsublings (arraylist S)()
+Public void setlevel (Tree Node, int hight)()
+Public void setposttree(Node tree, int posted))()
+Public void treepreordertree (Tree node)()
+Public void constructinvertedindex (Tree Node)()
+Public void Constructree(Element Node, Tree Node)()
+Public void lastdescendent (Tree Node)()

Node

-Array List <integer> alist
-Private String Trent
+Public void settent (string data)()
+Public string gettext()()
+Public Arraylist <integer> get Alist ()()
+Public void setalist (Arraylist <int> alist)()
+Public int Compareto(object argo)()

Similarity vector

-
+public void distance (patternindex)()

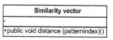

Bucket

-Public node
+Arraylist <node> listnodes()

Pattern Index

-linkedlist <node> alist
-Arraylist <integer> d
-double sim
+Public void addnode (node)()
+Public linkedlist <node> getalist()()
+public void setlistalist (linkedlist <node> alist)()

---

mapping

-Arraylist <double> mpsim
-Arraylist <string> checkp
-Arraylist <string> checkf
-Arraylist <double> totalsm
+Public <double> distance (arraylist<bucketb1>)()
+Public double nodesimilarity (node, node)()

Figure 7.3   Process class specification diagram.

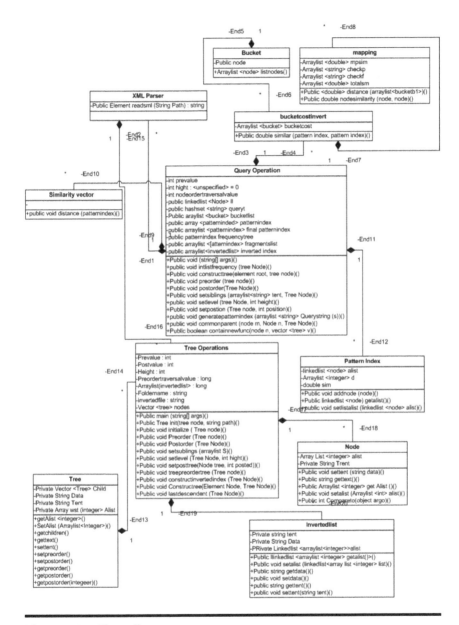

Figure 7.4   Class diagram.

Demo Screenshots

In the following passage, we show the screenshots from the prototype of causal reasoning based fault identification system. We use Java Beans to implement this infrastructure. Java Beans technology is the component architecture for the Java 2 Platform, Standard Edition (J2SE). Components (Java Beans) are reusable software programs that you can develop and assemble easily to create sophisticated applications. Java Beans technology is based on the Java Beans specification.

Figure 7.5 shows that we considered a part of query logs that is extracted from the log bank and found the related segments in the specification templates.

Figure 7.6 shows the similarity count with different fragments found in the specification repository.

Figure 7.7 shows the action of similarity measures and building of candidate matches to be forwarded to the policy engine.

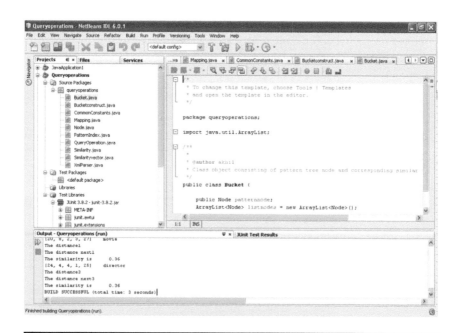

Figure 7.5   Screenshot: first algorithm implementation.

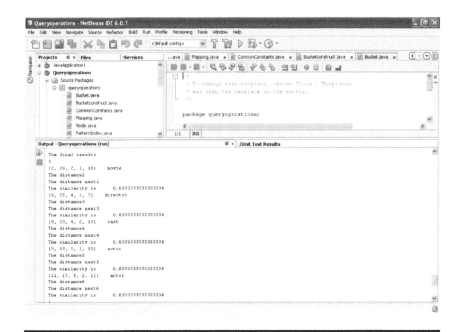

Figure 7.6   Screenshot: algorithm execution.

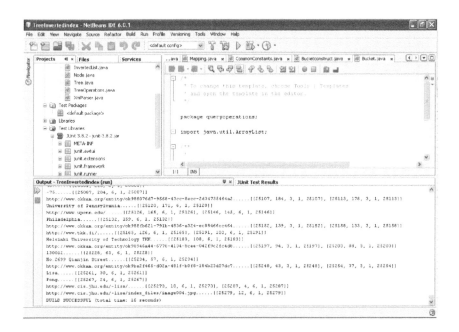

Figure 7.7   Screenshot: second phase execution.

Figure 7.8 demonstrates a test run of the policy engine. The case study of credit rule has been studied. The log from implementation classes of the case study has been extracted and after performing the causal reasoning based fault detection scheme, this experiment extracts the analysis file and generates a healing policy file from it for the client.

Figure 7.9 demonstrates the implementation experience of the policy engine. The case study of credit rule has been studied. The log from implementation classes of the case study has been extracted and after performing the causal reasoning based fault detection scheme, this experiment extracts the analysis file and generates a healing policy file from it for the client. Through these screenshots, we demonstrate that we have successfully tested and run the proposed scheme in a distributed computing environment.

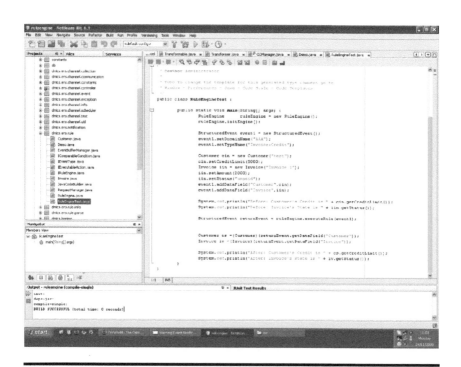

Figure 7.8   Test run of policy engine.

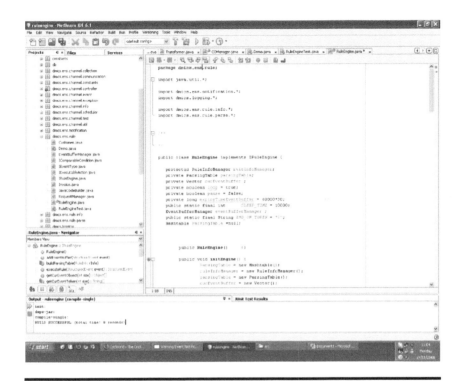

Figure 7.9   Policy engine implementation.

Reference

1. Knuth, D. E. *The Art of Computer Programming*, 3rd ed., Addison-Wesley, Reading, MA, 1997.

Chapter 8

Evaluation

In this chapter, we conduct various experiments in order to evaluate the performance of the fault detection scheme proposed in this book. The first experiment is performed to test the top K (a set of results returned after the first pass of the query parser) performance and precision of the relevant results of various algorithms proposed for fault detection in self-healing systems. We use TopX technology for this experiment.

TopX is a search engine for ranked retrieval of XML (and plain text) data. It supports a probabilistic-IR scoring model for full-text content conditions and tag-term combinations, path conditions for all XPath axes as exact or relaxable constraints, and ontology-based relaxation of terms and tag names as similarity conditions for ranked retrieval. For speeding up top-K queries, various techniques are employed: probabilistic models as efficient score predictors for a variant of the threshold algorithm, judicious scheduling of sequential access for scanning index lists and random access to compute full scores, incremental merging of index lists for on-demand, self-tuning query expansion, and a suite of specifically designed, precomputed indexes to evaluate structural path conditions.

Performance Graphs

Fault Detection and Traversal

The experiments we performed were on a 3-GHz dual Pentium PC with 4 GB of memory. Index lists were stored in an Oracle 11g server running on the same PC. Index lists were accessed in a multi-threaded mode and prefetched in batches of B = 200 tuples per index list and cached by the JDBC driver. The scan depth for invoking periodic queue rebuilds was synchronized with this prefetch size. The queue bound was set to Q = 500 candidates.

The dataset for this experiment was collected from the functional logs of the OKKAM project. The dataset consists of the events registered by various Java classes during their operations, every time they are invoked. This XML heap consists of a large file that is stored in the Oracle database. The Apache Tomcat server is used to emulate the client-server environment and load generate to generate queries that might be needed during the experiment. The Snowball stemmer from the Lucene project is used for indexing and parsed using the Oracle XML parser 11g.

We emulated the procedures proposed in Reference 1 as an application of fuzzy models, Reference 2 as an application of semiquantitative models, Reference 3 as an application of expert systems, Reference 4 as an application of knowledge-based systems, and Reference 5 as an application of artificial neural networks. The following results are obtained from our experiments.

Figure 8.1 shows that the performance of the scheme proposed in this book shows better performance than the others with which it is compared. There is a slight decrease in performance as the file source size increases, but the advantage of the proposed causal reasoning based scheme shows better performance.

The pattern-based systems generally have high response time, but as the size of the source file increases, they tend to

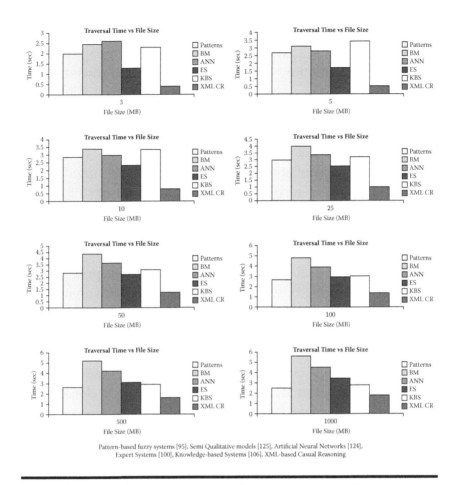

Figure 8.1   Fault identification time through traversal of increasing source size.

improve performance. This trend is observed as the *similar* instances show up in the activity log and the pattern-based systems develop activity patterns, experience-based, which make them use a longer response time but processing a relatively larger source file becomes more *cost effective* with pattern-based systems.

The semiquantitative models proved to be the costliest models tested in our experiment. On the other hand, the artificial neural networks (ANN) learn with time and their performance improves by 5% with the increase in size.

The ANNs are a good choice for extremely large storages with uniform complexity. Our experiment contains heterogeneous data (mapping after human users' unpredictable behavior) that makes ANNs take up a fair amount of processing time.

The expert systems, as the domain knowledge is fed to them beforehand, keep a consistent processing time throughout varying phases of source size. The knowledge-based systems pose an interesting proposition in the results of the experiment. Since our dataset is highly heterogeneous, every time the increase in source size is observed, the knowledge-based systems tend to show a linear increase in trend.

The causal reasoning based scheme that we propose in this book shows the best performance over highly varying datasets. In highly dynamic and heterogeneous situations, the scalability and efficiency are of essence. As the increase in size of the source takes place, there is no substantial increase in response time. This is proof of the fact that the presumption that we made after analytical study in the previous chapters was correct. Querying a very large database with a specific query and retrieving the desired results in the form of anomalies is the floor function of this research.

Resource Utilization

In the second experiment, we map study the resource utilization of each scheme under study. As we see the results in the following figures, ANN take the most amount of resources and with the increasing trend in the source size, the trend of increased use resources utilization continues. The Entity Manager (EM) starts with relatively higher resources. We observe that they show slight increase in resource utilization. The same goes for pattern-based systems, knowledge-based systems, and expert systems. They all show a steady increase in the resource utilization.

The causal reasoning based scheme proposed in this book works very efficiently and coincides with the other schemes in terms of resource utilization. In fact, it performs the best with the least amount of resources used (approximately 60 megabytes of resource file sizes). This trend shows that one does not need a high-end computing infrastructure to run the causal reasoning based scheme proposed in this book (Figure 8.2).

Generally, the causal reasoning based system needs a large amount of knowledge to cover various aspects within a context. In order to do this, it needs a very large memory and very fast processors or an efficient mechanism that could use its innate capabilities and make the case traversal less taxing on the resources. That is the main motivation behind our research, that is, to enable causal reasoning based systems with an inverted index and capabilities of XML technology so that they could perform to their capacities under relatively finite resources.

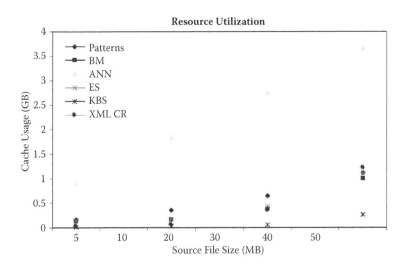

Figure 8.2   Resource utilization comparison.

Trends of Causal Reasoning Based Fault Detection

In this section, we study various trends of the causal reasoning based scheme independently.

Figure 8.3(a) shows that once the file source is converted into an inverted index, the size of the source does not have a substantial effect on the performance of the algorithms proposed. Figure8.3(b) shows a linear trend while constructing an inverted index from an increasing file source. Figure 8.3(c) shows that the performance of the causal reasoning based scheme was better than linear time until the eleventh node. Since the proposed algorithm is for small queries and large database source, we are satisfied with 11 nodes.

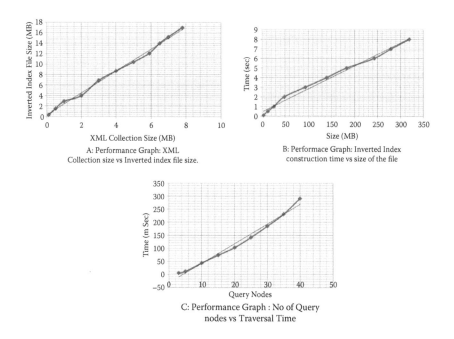

A: Performance Graph: XML
Collection size vs Inverted index file size.

B: Performace Graph: Inverted Index
construction time vs size of the file

C: Performance Graph : No of Query
nodes vs Traversal Time

Figure 8.3   Trends of casual reasoning based fault detection scheme.

References

1. Benouarets, M. and Dexter, A. On-line fault detection and diagnosis using fuzzy models, in IFAC Workshop, On-line Fault Detection and Supervision in the Chemical Process Industries, Newcastle, pp. 213–218, June 12–13, 1995.
2. De Kleer, J. and Brown, J. A qualitative physics based on confluences, *Artificial Intelligence*, 24:7–84, 1984.
3. Saludes, S., Corrales, A., Miguel, L., and Peran, J. A SOM and expert system based scheme for fault detection and isolation in a hydroelectric power station, in Proceedings Safeprocess 03, Washington, D.C., pp. 999–1004, 2003.
4. Rengaswamy, R. and Venkatasubramanian, V. An integrated framework for process monitoring, diagnosis, and control using knowledge-based systems and neural networks, in IFAC Symposium, On-line fault detection and supervision in the chemical process industries, Newark, DE, pp. 49–55, April 22–24, 2007.
5. Can, C. and Danai, K. Fault diagnosis with a model-based recurrent neural network, in Proceedings Safeprocess 03, Washington, D.C., pp. 699–704, 2003.

Chapter 9

Contributions

Contributions are made in three broad areas. The first consists of conceptual ideas underpinning the work. The second consists of the architectures created in the course of the dissertation. The third set of contributions is the lessons learned from the qualitative and quantitative evaluation of these architectures.

Each of these areas is discussed in the following sections.

Conceptual Contributions

Causal Reasoning Based Fault Identification

Causal reasoning (CR) is rather an unusual candidate about which we seek to find the answers on fault identification in self-healing systems. However, our implementation of policy engine and multifaceted rule manipulation techniques posed an interesting proposition as a lot of background knowledge and exception-handling scenarios enable the use of CR. We implemented it and the results were promising.

Self-Management Functional Hierarchy

We identify the role of self-healing, which is mostly misunderstood among contemporary systems, and consider that the complexity among the autonomic systems can be reduced through rearrangement of self-* system and redefining their roles. The identification of some logical problems especially in fault mapping, functional classification, and categorization are among the major motivations of our work.

Several network management solutions proposed in References 1 through 4 are confined strictly to their respective domains, that is, either mesh network or MANETs. Self-management architecture is proposed in Reference 5 for u-Zone networks. In what already is a very limited amount of published work, we have identified the following questions still to be answered:

- If self-healing is one of the FCAPS (Fault, Configuration, Accounting/Administration, Performance, Security) functions, then what is the physical location of self-healing functions? Should they reside on the gateway or at the client end?
- How does the control, information, etc. flow from one function to another? Especially, how do self-healing functions interact with other functions?
- What are the calling signatures of self-healing functions? If self-healing functions are fault-removing functions, then what are the functions of fault management functions?
- Are these sub-functions functionally independent? If yes, then is there evidence of redundancy and if not then how can self-healing be thought of as an independent entity? In other words, what is the true functionality definition of self-healing?
- If the self-management entity faces a management problem, how should it be tackled?

The conventional methods of system evaluation should not be used for autonomic systems. The sum of complexities of all the components of a system normally shows the system complexity. In a ubiquitous network like a u-Zone network where the system components are entirely distributed, it would be erroneous to include non-management related modules in computational complexity calculations of self-management modules. Moreover, we assume that the executable components are of atomic nature and perform a certain "basic" function. We generate a self-healing policy from selecting, rearranging, and calling these components. According to References 6 and 7, the executable components (in our case, plug-ins) can be classified as irreducibly complex. Following this derivation, we conclude that the executable components, which are provided by a third-party vendor, should not be included in the computational complexity of self-healing systems.

These reservations encourage us to propose software architecture that follows the principles of self-growing, self-managing, and autonomic in nature, which is the very spirit of our project.[8] Unlike numerous application domains reported in the literature,[9–30] we apply our findings[31] in network management where little work has been done, to the best of our knowledge.

The autonomic self-management in hybrid networks is a relatively new area of research. In Reference 5, the authors propose an autonomic self-management framework for hybrid networks. Our approach is different from their work in basic understanding of the functionality of self-management functions. We argue that the self-management functions do not stem from one main set; rather, they are categorized in such a way that they form on-demand functions and some functions are always-on/pervasive functions.[32]

We propose that the context-awareness and self-optimization should be an always-on category and the others as on-demand functions. This approach is very useful in hybrid environments where there are clients of various batteries and computing powers, memories, etc. and a self-management framework might prove to be too much middleware. The normal functionality model (NFM) regulates the usage of self-management functions according to computing ability of the client. This gives the client level local self-management. The healing policies come in two parts: condition and action. The condition parts are extracted from the service request that the client sends. The analyzer analyzes the service requests through the procedures discussed in References 33 through 37.

Multifaceted Policy Engine Methodology

As business applications become complex and changeable, a rule-based mechanism is needed for automatic adaptive computing as well as trustworthy and easy maintenance. For this purpose, we propose a compilation-based policy engine that can easily express business rules in Java codes. It does not need additional script language for expressing rules. It can create and execute condition and action objects at run time. Moreover, it can use existing libraries for condition or action codes of rules (i.e., String, Number, and Logical Expression) so that not only can it express complex condition or action statements, but also it can easily integrate the existing systems developed in Java. The compilation-based policy engine proposed in this chapter shows better performance than JSR-94, a generally used interpretation-based policy engine. According to our experiments, the proposed policy engine processes 245 more transactions per second than does JSR-94. We intend to test the performance of the policy engine proposed in this research with different weights and in different conditions. This will not only give us a better idea about the working capacity of the outcome of this research, but also it will give a clear application area for this policy engine. Moreover, we intend to develop a GUI that could assist users who have limited knowledge of Java in operating with this policy engine.

Healing Policy Generation

The results from the fault detection module are fed to the policy engine and a healing policy is generated for self-healing at a remote client. This healing policy uses the multifaceted rules that cover all aspects related to a fault. The aggressive fault debugging gives better results in a u-Zone network.

The mapping of software specification with the application logs pinpoints the exact location of the fault. This fault localization technique helps us to apply our aggressive scripts in order to remove the bugs.

Architectures

AHSEN

In hybrid wireless networks, there are many different kinds of devices attached to the network. They vary from each other in the bases of their power, performance, etc. One of the characteristics not present in the related literature is the separate classification of the client and the gateway architectures.

The NFM is a device-dependent ontology that is downloaded on the device at the network configuration level. It provides a mobile user with an initial default profile at the gateway level and device level functionality control at the user level. The updating of the profile is done by user activities during its active session with the gateway. It contains the normal range of functional parameters of the device, services environment, and network, which allows prompt anomaly detection. The SMF constantly traps the user activities and sends them to the SMF at the gateway. The SMF at the gateway directs the trap requests to the context manager who updates the related profile of the user. The changes in service pool, trust manager, and policy manger are reported to the context manager. The context manager consists of a lightweight directory access protocol

(LDAP) directory that saves its sessions after some time intervals in the gateway directory. The LDAP directory servers store their information hierarchically. The hierarchy provides a method for logically grouping (and sub-grouping) certain items together. It provides scalability and security, and resolves many data integrity issues. The policy manager (PM) and service manager (SM) follow the same registry-based approach to enlist their resources. The presence of NFM provides the decision-based reporting, unlike the ever-present SNMP. The trust manager uses the reputation-based trust management scheme in public key certificates.[33] The trust is typically decided on the trustee's reputation. The trust-based access relies on "soft" to mitigating risks involved in interacting and collaborating with unknown and potentially malicious users. Due to reiterative reasons, different kinds of repositories located at the gateway are not specified in the architectural diagram.

Policy Engine

We proposed a compilation-based dynamically adjustable policy engine that is used for rich rule expression and performance enhancement. Because of immense complications among business rules, it is very costly to manage the change in modern-day business systems. We propose to use Java to create/modify rules instead of scripting languages. Java gives us the facility of standardized syntax and eliminates the natural language ambiguities from the program. Moreover, with the technical support present for Java worldwide in the form of an IT workforce, Java is the programming language of choice. We propose that separating the condition from action part during run time makes rule notification easier and quicker. According to experimental results, the proposed dynamically adjustable policy engine shows promising results when compared with contemporary script-based solutions. The applications of the proposed systems can be found in autonomic self-management systems among our research.

Software Prototypes

We have developed a CR-based fault detection scheme and policy engine modules that are a part of our Autonomic Healing-based Self-Management Engine (AHSEN).
Some other contributions are:

■ Identification of irreducible complexity as one of the problem areas in real-life, widespread application of autonomic systems.
■ Propose a self-growing distributed framework for self-management that considers local, medium, and global level self-management.
■ Context management and optimization of a system pervasively with on-demand fault management.

References

1. Burke, R. *Network Management. Concepts and Practice: A Hands-On Approach*, Pearson Education, Inc., Upper Saddle River, NJ, 2004.
2. Oh, M. Network management agent allocation scheme in mesh networks, *Communications Letters*, 7(12):601–603, 2003.
3. Kishi, Y., Tabata, K., Kitahara, T., Imagawa, Y., Idoue, A., and Nomoto, S. Implementation of the integrated network and link control functions for multi-hop mesh networks in broadband fixed wireless access systems, Radio and Wireless Conference, pp. 43–46, Sept. 19–22, 2004.
4. Yong-Lin, S., DeYuan, G., Jin, P., and PuBing, S. A mobile agent and policy-based network management architecture, Proceedings, Fifth International Conference on Computational Intelligence and Multimedia Applications, ICCIMA 2003, pp. 177–181, Sept. 27–30, 2003.
5. Chaudhry, S. A., Akbar, A. H., Kim, K.-H., Hong, S.-K., and Yoon, W.-S. *HYWINMARC: An Autonomic Management Architecture for Hybrid Wireless Networks*, Network Centric Ubiquitous Systems, 2006.

6. Behe, M. J. *Darwin's Black Box: The Biochemical Challenges to Evolution*, Touchstone, New York, 1996.
7. Irreducible Complexity Revisited. Available at www.iscid.org. Accessed November 12, 2006.
8. u-Frontier: Ubiquitous Korea. Available at www.uauto.net. Accessed on November 12, 2006.
9. Boonma, P. and Suzuki, J. BiSNET: A biologically inspired middleware architecture for self-managing wireless sensor networks, *Comput. Networks,* 51(16):4599–4616, 2007.
10. Krena, B., Letko, Z., Tzoref, R., Ur, S., and Vojnar, T. Healing data races on-the-fly, in Proceedings of the 2007 Workshop on Parallel and Distributed Systems: Testing and Debugging (PADTAD 2007), London, England, July 9, 2007.
11. Shehory, O. A self-healing approach to designing and deploying complex, distributed and concurrent software systems. *Lecture Notes in AI*, Vol. 4411, Bordini, R., Dastani, M., Dix, J., and El Fallah Seghrouchni, A., Eds., Springer-Verlag, pp. 3–11, 2006.
12. Breitgand, D., Goldstein, M., Henis, E., Shehory, O., and Weinsberg, Y. PANACEA towards a self-healing development framework, *Integrated Network Management*, 169–178, 2007.
13. Trumler, W., Ehrig, J., Pietzowski, A., Satzger, B., and Ungerer, T. A distributed self-healing data store, *ATC*, 458–467, 2007.
14. Anglano, C. and Montani, S. Achieving self-healing in service delivery software systems by means of case-based reasoning. *Applied Intelligence*. To appear.
15. Zhang, X., Dragffy, G., Pipe, A. G., Gunton, N., and Zhu, Q. M. A reconfigurable self-healing embryonic cell architecture, *Engineering of Reconfigurable Systems and Algorithms*, 134–140, 2003.
16. Trumler, W., Helbig, M., Pietzowski, A., Satzger, B., and Ungerer, T. Self-configuration and self-healing in AUTOSAR, 14th Asia Pacific Automotive Engineering Conference (APAC-14), August 5–8, 2007, Hollywood, CA.
17. Ghosh, D., Sharman, R., Raghav Rao, H., and Upadhyaya, S. Self-healing systems—survey and synthesis, *Decis. Support Syst.,* 42(4):2164–2185, 2007.
18. Bokareva, T., Bulusu, N., and Jha, S. SASHA: towards a self-healing hybrid sensor network architecture, in Proceedings of the 2nd IEEE International Workshop on Embedded Networked Sensors (EmNetS-II), Sydney, Australia, May 2005.

19. Trumler, W., Petzold, J., Bagci, F., and Ungerer, T. AMUN: an autonomic middleware for the Smart Doorplate Project, *Personal Ubiquitous Comput.*, 10(1):7–11, 2005.
20. Sajjad, A., Jameel, H., Kalim, U., Han, S. M., Lee, Y.-K., and Lee, S. AutoMAGI—an autonomic middleware for enabling mobile access to grid infrastructure, Joint International Conference on Autonomic and Autonomous Systems and International Conference on Networking and Services (icas-icns'05), p. 73, 2005.
21. Siewert, S. and Pfeffer, Z. An embedded real-time autonomic architecture, IEEE Denver Technical Conference, April 2005.
22. Wile, D. and Egyed, A. An externalized infrastructure for self-healing systems, Proceedings of the 4th Working IEEE/IFIP Conference on Software Architecture (WICSA), pp. 285–288, Olso, Norway, June 2004.
23. Can, W., Yang, L., and Jianjun, B. A biological formal architecture of self-healing system, *SMC*, (6):5537–5541, 2004.
24. Shen, C., Pesch, D., and Irvine, J. A framework for self-management of hybrid wireless networks using autonomic computing principles, in Proceedings of the 3rd Annual Communication Networks and Services Research Conference (Cnsr'05), pp. 261–266, Washington, D.C., May 16–18, 2005.
25. Liu, P. ITDB: an attack self-healing database system prototype, *DISCEX*, 2:131–133, 2003.
26. Sharmin, M., Ahmed, S., and Ahamed, S. I. MARKS (middleware adaptability for resource discovery, knowledge usability and self-healing) for mobile devices of pervasive computing environments, Proceedings of the Third International Conference on Information Technology: New Generations (ITNG 2006), pp. 306–313, Las Vegas, NV, April 2006.
27. Wang, F. and Li, F.-Z. The design of an autonomic computing model and the algorithm for decision-making, *GrC*, 270–273, 2005.
28. Gangadhar, D.K. Meta dynamic states for self healing autonomic computing systems, 2005 IEEE International Conference on Systems, Man and Cybernetics, pp. 39–46, Oct. 10–12, 2005.
29. Grishikashvili, E., Pereira, R., and Taleb-Bendiab, A. Performance evaluation for self-healing distributed services, Proceedings of the 11th International Conference on Parallel and Distributed Systems—Workshops (Icpads'05), pp. 135–139, Washington, D.C., July 20–22, 2005.

30. Zenmyo, T., Yoshida, H., and Kimura, T. A self-healing technique using reusable component-level operation knowledge, *Cluster Computing*, 10(4):385–394, 2007.
31. Chaudhry, J. A. and Park, S.-K. Some enabling technologies for ubiquitous systems, *Journal of Computer Science*, 2(8):627–633, 2006.
32. Chaudhry, J. A. and Park, S. Using artificial immune systems for self-healing in hybrid networks, in *Encyclopedia of Multimedia Technology and Networking*, Idea Group Inc., U.S.A., 2006.
33. Garfinkel, S. *PGP: Pretty Good Privacy*, O'Reily & Associates Inc., 1995.
34. Trumler, W., Petzold, J., Bagci, F., and Ungerer, T. AMUN–Autonomic Middleware for Ubiquitious eNvironments Applied to the Smart Doorplate Project, International Conference on Autonomic Computing (ICAC-04), New York, May 17–18, 2004.
35. Gao, J., Kar, G., and Kermani, P. Approaches to building self-healing systems using dependency analysis, Network Operations and Management Symposium, April 19–23, 2004, IEEE/IFIP 1:119–132, 2004.
36. Chaudhry, J. and Park, S. On seamless service delivery, the 2nd International Conference on Natural Computation (ICNC'06) and the 3rd International Conference on Fuzzy Systems and Knowledge Discovery (FSKD'06), 2006.
37. Chaudhry, J. A. and Park, S. AHSEN—Autonomic Healing-based Self-management Engine for Network management in hybrid networks, the Second International Conference on Grid and Pervasive Computing (GPC07), 2007.

Chapter 10

Conclusion

In an on-demand business, information technology (IT) professionals must strengthen the responsiveness and resiliency of service delivery—improving quality of service—while reducing the total cost of ownership (TCO) of their operating environments. A slowdown in Moore's law is in near sight and seen as the main obstacle to further progress in the IT industry. Rather, it is the industry's exploitation of the technologies that have been developed in the wake of Moore's law that has led us to the verge of a complexity crisis. Software developers have fully exploited a 4 to 6 order-of-magnitude increase in computational power—producing ever more sophisticated software applications and environments. Exponential growth has occurred in the number and variety of systems and components. The value of database technology and the Internet have fueled significant growth in storage subsystems, which now are capable of holding petabytes of structured and unstructured information. Networks have interconnected, distributed, heterogeneous systems. Our information society has created unpredictable and highly variable workloads for these networked systems. Moreover, these increasingly valuable, complex systems require highly skilled IT professionals to install, configure, operate, tune, and maintain them.

The concept of building a municipal broadband wireless mesh network to cover an entire city, metropolitan area, or county has been broadly examined in recent years. Literally hundreds of municipalities worldwide have begun to deploy such networks, based on Wi-Fi, public-safety band, and WiMAX technology, and thousands more are in the planning or proposal stages.

Most observers recognize that such networks can offer a myriad of potential benefits. A municipal wireless network can provide inexpensive, ubiquitous Internet access to economically disadvantaged citizens and public schools, an initiative referred to as "digital inclusion" or "bridging the digital divide." It can spur economic development by making the city more attractive to tourists and businesses. A citywide broadband wireless network can support a variety of mobile applications that greatly improve the efficiency and responsiveness of public safety and municipal workers. Wireless video surveillance can be used to improve public safety. Ubiquitous wireless coverage can help automate, streamline, and reduce the costs of delivering basic city services like utility metering and parking enforcement.

However, a municipal wireless network is a large-scale, high-stakes project that demands significant political will, capital investment, and commitment of personnel resources. Careful strategic planning, a clear-eyed assessment of the technical challenges, and consideration of funding options are all essential to making a citywide wireless network a success for municipal government, its employees, businesses, visitors, and citizens. Municipalities must find a way to build a broadband wireless network that makes economic sense while addressing the current and future needs of the broadest range of public access, public works, and public safety applications.

Self-managing capabilities in a system accomplish their functions by taking an appropriate action based on one or more situations that they sense in the environment. The function of any autonomic capability is a control loop that collects details from the system and acts accordingly. In a self-managing autonomic environment, system components— from hardware (such as storage units, desktop computers, and servers) to software (such as operating systems, middleware, and business applications)—can include embedded control loop functionality. In a ubiquitous network, like a u-Zone network where the system components are entirely distributed, it would be erroneous to include non-management related modules in computational complexity calculations of self-management modules. Moreover, we assume that the executable components are of atomic nature and perform a certain "basic" function. We generate a self-healing policy from selecting, rearranging, and calling these components. According to References 1 and 2, the executable components (in our case, plug-ins) can be classified as irreducibly complex. Following this derivation, we conclude that the executable components, which are provided by a third-party vendor, should not be included in the computational complexity of self-healing systems.

In the presence of computational complexity, one function that should be pinpointed and should take the least analytical time during the working lifetime of a system is fault identification. Our study and experience of implementation of a self-healing system has made us use various techniques, but all of them seem to take a lot of resources and time. This research targets the fault identification techniques that should be used in a self-healing based system. The use of causal reasoning (CR) in the implementation of the fault identification study substantially improves the performance of the fault detection. We enlist and analyze various fault detection schemes and present our previous theoretical research as a case study and implement the CR-based fault identification scheme and demonstrate better performance through empirical results. The empirical results obtained from our experiments show a marked improvement in activity log traversal size and hence fault identification. We demonstrate that the proposed scheme works better than equiquantitive models, artificial neural networks, expert systems, knowledge-based systems, etc.

The following areas of research are left for future work.

- Context-Specific Applications: The context in a software program is important in fault detection. Now we are relying on the software specifications that are extracted during the software development lifecycle. In order to broaden the applicability of self-healing based systems, one needs to explore more varying contexts and applications.
- Uncertainty Management: The uncertainty has always been among the forefront of fault detection in autonomic systems. We have yet to measure the uncertainty trends in AHSEN, as it will take time to populate the CR knowledge and accuracy of the healing policy artifacts.
- Self-Healing When There Is Not Enough Knowledge: Some nascent and domain-specific knowledge is embedded into AHSEN, but in the future knowledge mining crawler should be used to populate the CR base. An example of such a knowledge crawler is OKKAM crawlers.
- Application Complexity: As the application domains increase, so will the complexity of the domain knowledge and thus the application complexity of AHSEN.

References

1. Behe, M. J. *Darwin's Black Box: The Biochemical Challenges to Evolution*, Touchstone, New York, 1996.
2. Irreducible Complexity Revisited. Available at www.iscid.org. Accessed November 12, 2006.

Index